THE ROBOT AND AUTOMATION ALMANAC 2019

FI THE FUTURIST INSTITUTE

Edited by Jason Schenker
Chairman of The Futurist Institute

THE ROBOT AND AUTOMATION ALMANAC - 2019

THE FUTURIST INSTITUTE

EDITED BY JASON SCHENKER

ISBN: 978-1-946197-14-6 *Paperback*
978-1-946197-15-3 *Ebook*

PRESTIGE
PROFESSIONAL PUBLISHING

For automation and robotics futurists everywhere.

THE FUTURIST INSTITUTE

CONTENTS

CONTENTS

CONTENTS

PREFACE

From The Futurist Institute

On behalf of The Futurist Institute, I want to acknowledge and thank all of the contributors to *The Robot and Automation Almanac - 2019*. We have such an amazing cohort of contributors in the fields of robots and automation this year.

Thank you to Samuel Alexandersson, Djamila Amimer, Fred van Beuningen, Keith Blanchet, Jeff Burnstein, Lorenzo Carver, Brian La Cour, Craig Fuller, Jason Gouw, Rob Handfield, Robert Huschka, Maciej Lisiecki, Jayesh Mehta, Russell Nickerson, David Schwebel, Daniel Stanton, Kaleb Steinhauer, Daniel Theobald, Michael Walton, and Bruce Welty. I am truly humbled by this list of amazing authors!

The contributions of these leaders will help shape the vision for robots and automation in the year to come — and beyond. Their words and expectations for "the big thing" in the year ahead have created an important vision of the future. And without them, this book would never have happened. I believe you will find the visions of our authors for the year ahead to be exciting and valuable!

I also want to thank parties who have not contributed to this work but who have helped out by providing recommendations for potential contributors, including MHI — the Material Handling Industry group — with which my financial market research firm, Prestige Economics, has a long-standing relationship. I am also grateful to Martin Buehler, who introduced me to several of the authors in this work. They are experts, and without Martin's introductions, this book would not be as rich as it is.

Finally, I need to thank the individuals who have provided support and feedback to this project, especially Nawfal Patel, who helped managed the process. Of course, I am also grateful to my other colleagues at Prestige Economics and The Futurist Institute who helped me bring this book to fruition. And, of course, Kerry Ellis has my gratitude for designing such a wonderful cover for the almanac.

It is my personal pleasure to have been involved in this undertaking.

The robotics and automation industries are still very new, but they are becoming more important. And it is with great pleasure that The Futurist Institute is able to support the companies in these industries with this almanac. It is also our pleasure to work alongside the long-established thought leadership and conferences that have been so important for the development of robotics and automation, including A3 and the RoboBusiness conference.

The Robot and Automation Almanac represents a written gathering of thought leaders, but RoboBusiness is an amazing in-person experience. It has been formative for my own transition from economist to futurist — and for the very founding of The Futurist Institute.

I founded The Futurist Institute to help strategists, analysts, and executives become futurists in October 2016; we recently began our third year of operations. With the inaugural edition of *The Robot and Automation Almanac* in 2018, The Futurist Institute created an invaluable tool for professionals, individuals, and investors seeking to understand the implications of robots and automation for their personal, professional, and investing lives.

We have similarly high expectations for how *The Robot and Automation Almanac - 2019* will presage changes in the year ahead.

The transition from the information age into the automation age is underway. Thank you for being a part of this dynamic shift.

Welcome to the future! ~

Jason Schenker
Chairman of The Futurist Institute
Editor of *The Robot and Automation Almanac*

The Year Ahead and Beyond

The main theme of *The Robot and Automation Almanac - 2018* was that 2018 was going to be a big year for robots and automation. It was going to be a critical year of transition, of momentum, of activity, of investment, and of technological potential. And we were going to see robots leave the factories and be everywhere.

And that's pretty much how 2018 played out.

We are standing on the edge of the age of automation. What's next for robots and automation in the year ahead? This is the question that our contributors have answered in this book.

Robots and automation received a lot of media attention in 2017 and 2018. This visibility and emergence in the technological *Zeitgeist* is a precursor of what is likely to happen in 2019. Our authors have presented a view of a world in which robots and automation will make progress. And it will be a world in which AI and computational processing power will be increasingly important.

Of course, there are also some challenges ahead for 2019.

It may be a year in which robots are forced to "slow down," as Michael Walton points out in his essay, because of a need to wait for software to catch up with the robots. As many people have begun to see and realize, robots are getting ready to have a step change impact on society. But the breakneck pace of development necessitates some potential retrenching.

Another potential challenge in the year ahead could come from some more challenging business dynamics — dynamics that could, according to Bruce Welty's essay, bring about a "reckoning" for robotics. Economic and financial market risk dynamics going into 2019 are very different than those going into 2018. After all, we now have higher interest rates and trade risks that did not exist a year ago, and there are elevated concerns about the potential for tighter and stricter foreign investment allowed into the United States as part of stricter oversight and control by the Committee on Foreign Investment in the United States (CFIUS).

In the essays in this book, you will see a few different threads among the different authors.

Last year's theme was that 2018 was a year of ramping up activity. And we divided *The Robot and Automation Almanac - 2018* into two parts: "Robots You See" and "Robots You Don't See."

But as our contributors predicted, robots started popping up everywhere, to the point that this split of topics doesn't make sense in 2019, Essentially, almost all robots are now being seen.

That's why this year, for the 2019 almanac, we have divided the book into four different sections:

— Outlook for Robotics, Automation, and AI in 2019
— Socializing Robots
— Robots in Industry
— Robots in the Supply Chain

In the first section, which is about the **Outlook**, we look at the macro factors — some of which I have just alluded to — that are likely to drive robotics, automation, and AI in the year ahead. There are some big needs on the software side of robots in order to maximize the benefit they bring; Djamila Amimer provides a detailed list of a number of ways that AI will become important in the year ahead — and beyond. Plus, Maciej Lisiecki offers a broad view on robotics growth areas in 2019.

Additionally, I am excited that Austin-based Brian La Cour contributed an essay on the near-term outlook for quantum computing, "quantum supremacy," and his own project to create emulated quantum computing, to get computational benefits at room temperature.

The second section of the almanac is about **Socializing Robots**. Jason Gouw and Russell Nickerson contributed very powerful essays about this subject. Plus, Lorenzo Carver offered an outlook for voicebots in 2019, and Daniel Theobald and Keith Blanchet focus on the economic value add robots bring — and the need to overcome the aversion some people have to working with them. Because robots are here, and we need to leverage them.

In the third section of the almanac, we look at **Robots in Industry**, and our contributors offer insight into what to expect in the year ahead for robots, automation, and AI in some very specific industries. The essays include examinations of transportation (by Craig Fuller), smart cities (by Fred van Beuningen), construction (by Kaleb Steinhauer), finance (by Rob Handfield), and the overall concept of smart industries (by Jeff Burnstein & Robert Huschka). These contributions get deep into specific industries and offer a glimpse into the myriad of ways that robots, automation, and AI are likely to have an impact in 2019.

In the fourth and final section of *The Robot and Automation Almanac - 2019*, we look at the hot topic of **Robots in the Supply Chain**. The section kicks off with an overview of potential developments in the supply chain from Daniel Stanton, whose moniker online is "Mr. Supply Chain." This essay is followed by Samuel Alexandersson's focus on warehouses riding the wave of the fourth industrial revolution.

The supply chain section also includes an essay from David Schwebel, which offers some tremendously accessible insights through common examples that make it easy to go deep on the concept of hyperlocal supply chains. The final contribution to the book is an introspective essay from Jayesh Mehta on skill gaps.

In addition to changing the format of *The Robot and Automation Almanac - 2019* to include four sections of contributions, we also included more images of robots and automation activities because we got tremendously positive feedback about pictures last year.

In addition to their contributions, we are also grateful for all of the images our contributors granted us permission to use in this book.

The history of robots and automation is being written now.

In this book.

When people look back a decade from now, they will want to know how robots and automation evolved. They will want to understand how robots and AI became more integrated.

They will wish they had known more when it was happening.

They will wish they had read *The Robot and Automation Almanac* as it was happening. ~

Jason Schenker
Chairman of The Futurist Institute
Editor of *The Robot and Automation Almanac*

Outlook for Robotics, Automation, and AI in 2019

FI THE FUTURIST INSTITUTE

Vectors of Robotics Advancement in 2019

Michael Walton

- Microsoft Industry Solutions Executive -

There are three vectors that will be important for robotics, automation, and AI development in 2019.

These three factors are:
1. Rapid advancement in robotics, which is ongoing.
2. AI with a robotics intercept, which is on track.
3. Old stuff that needs to be pulled into an ever-evolving high-tech system.

Vector 1 — Rapid Robotics Development

We need to consider each of the three factors as we think about the future of robotics in the year ahead. And the advancement of robotics is likely to be strong and ongoing in manufacturing, logistics, and supply chain industries. This part of robotics is humming along.

The big picture developments and surprises are likely to come from the other two vectors, which are applying AI to robots and what to do with the old junk. These will be critical drivers of robotics dynamics on the year ahead.

Vector 2 — AI and Robots

When most people think of robots and AI, they're just thinking about the tasks the robot has to do. But when I talk about AI in bots, there are two ways I think about it: the way the bots do their jobs and the way the bots are maintained so they can do their jobs.

First, let's talk about AI in robots. And let me give you an example. I've been working with a company that does packaging and product inspection. And they do about 1 billion pick and pack placements annually. There is no AI currently built in, and the scrap rate is 4.8 percent.

With AI, they could be doing predictive error identification earlier in the process. And that could reduce their scrap rate to a much lower 1 percent. The economic value add and ROI could be massive by moving to AI robots.

Another area where using AI with robots creates value in task-based operations is by integrating bot activity. This is where you have robots working together, building AI, and there is a compounding impact that can outpace humans.

This kind of integrated learning structure in an AI robot system also seems particularly attractive recently because of a very tight US labor market. This is just another way that AI use in bot operations can solve a critical corporate problem.

But while adding AI to bot operations can bring tremendous value, there is also a need to add AI outside of the bots. This is the second area where AI and bots can be integrated to create value. And it may be even more valuable than having robots with AI in their operations.

This is almost a meta concept of AI use with bots because this could include predictive analytics on the maintenance of bots. This would help prevent downtime. And that is absolutely essential — especially for bots working under increasingly intense supply chain and e-commerce demands.

One of the important areas of predictive maintenance with bots is the push toward a hard swap. This allows for self-placement of a bot so that there is no operational downtime. And its flawless execution requires an integration of robots with AI.

But the integration of robots and AI can take time.

Something that is both interesting and surprising is that companies that have invested heavily in robots are now trying hard to slow it down in order to integrate their robot use with AI. Because this improves efficiency, adds greater value, and ultimately increases profits.

Vector 3 — AI and Old Stuff

One of the big developments in robots and AI we see in the year ahead is the increased attempt to integrate existing equipment, machinery, and operations with the newest technology. And this is where companies often struggle and need to slow down their AI and robotics initiatives. After all, you need the entire operation to work in an integrated fashion. And this means retrofitting old stuff.

A lot of this is being done with Internet of Things (IoT) sensors. About a year and a half ago, there were about 1,000 IoT sensors. But today, the number and type of sensors is boundless. There are innumerable IoT sensors for temperature, pressure, vibrations, leveling, GPS — you name it!

Using these sensors to retrofit old stuff — equipment, machinery, and operations — is easier, cheaper, and more possible now than ever before. And this could actually lead to even more use cases for robots and AI. This is especially true because not just sensors, but robots themselves, are also getting cheaper. And AI capabilities are improving so fast.

Challenges Ahead — Robots and AI Sold Separately

Currently, the biggest challenge to robot and AI use and integration isn't the cost of the robots and the sensors. Nor is the biggest challenge preventative maintenance or operational integration with old stuff.

The biggest challenge is how AI and bots are sold. Right now, machinery and equipment are usually purchased from a dealer. But the software and AI are not sold by the dealer. They are an add-on.

Essentially, equipment and AI are sold separately. This lack of integration creates delays, budget problems, and other implementation issues. They should be sold together, but they generally are not.

This means that manufacturers are held hostage by their dealers. And until robots and AI are sold together — until you can buy equipment and software bundled together — companies may struggle to achieve the efficiency, optimization, and ROI that robotics, automation, and AI can bring to a facility.

While 2019 is likely to be a big year for robot development and AI improvements, the biggest change to watch for is whether equipment and machinery can be sold with the AI software that creates the greatest value. It needs to happen. Hopefully, 2019 will be the year when it does. ~

Michael Walton is a Director and Industry Solutions Executive at Microsoft, where he is helping discrete manufacturing companies to digitally transform.

A Coming Robotics Reckoning: There Will be Big Winners and Big Losers

Bruce Welty

- Chairman of Locus Robotics, Inventor, Innovator -

Robots, Automation, and AI have made significant strides in recent years. But sometimes things have to take a step back before they can move ahead.

And 2019 could be a year where robotics takes a step back. It could be a year of reckoning for robotics.

Any why is a reckoning for robotics coming?

Well, it isn't because robotics is unimportant. In fact, robots have become a critical part of the supply chain and manufacturing industries. And they will become even more important in coming years.

A reckoning is coming because robotics needs to see a separation of the wheat from the chaff.

The economic and societal potential for robots is massive. And this has fueled some hype-driven investments. So now, it's time to ask: how many ideas will make it?

There have been a lot of great ideas that just go bust. Look at Jibo, the company that created a social robot. It was a great idea. But now, they're out of business.

And they won't be the last.

In other areas of robotics, human-looking robots have made tremendous strides in terms of increased capabilities and the ability to manage gravity. Now they not only walk, but what they can do looks increasingly like parkour. But just because you can do something doesn't mean there will be real economic value in it.

And maybe a bipedal humanoid robot isn't the right choice because it may not have the greatest economic value add or investor ROI.

Don't get me wrong. Robots that look like humans and can move around with the ability of a ninja are cool. And the same could be said about self-driving vehicles. But bipedal robots and self-driving cars have one critical thing in common that will plague their development with delays: they operate in a three-dimensional world. And that is a world filled with innumerable edge cases.

But while minimum viable products that impress investors are on our doorstep, truly commercially viable products may be a long way off.

And that is where the focus is likely to increasingly be: on commercialization, practical use cases, and return on investment. Because without revenue, robotics development is just a research project, and not a business.

Robots need a reason to exist beyond just being cool. Especially with increasingly more discerning investment dollars on the line. And this is certainly likely to be the case, given higher interest rates and a tightening of access to foreign investment and acquisition dollars due to changes at CFIUS, which is the Committee on Foreign Investment in the United States.

We are due for a robotics reckoning, and these factors present elevated risks.

But while there will be some losers ahead, there will also be big winners.

Robotics companies that are likely to excel in this kind of environment are those that operate in two-dimensional space, like robots in the supply chain that are working to fulfill the promise of e-commerce in warehouses and distribution centers. They move along vertical and horizontal axes, which reduces their edge cases and allows for simplicity that brings true economic value.

Of course, these aren't sexy robots that get the headlines, but their economic impact is nontrivial. Keeping robots operating in two-dimensional space and applying them to tasks that humans are bad at is the kind of simplicity with economic viability that makes me optimistic about my company and our industry as a whole.

Of course, there are other robotics use cases with tremendous value but less hype. These include robots that can perform dangerous tasks for humans, including in the military, underwater, and certain industrial jobs.

These companies have use cases of true value.

As I look ahead to 2019, companies with poor use cases and innumerable edge cases are particularly at risk. And this forces me to invoke an important aphorism that gets lobbed around the technology world: "You gotta have growth. You gotta have profits. But you can't have neither."

Because more discerning investment dollars mean that companies with neither growth nor profits will find themselves swept away by the reckoning ahead. ~

Bruce Welty is the Chairman at Locus Robotics Corporation. Locus Robotics provides an autonomous robot that streamlines warehouses and distribution centers to increase the efficiency of the fulfillment industry.

Source: Locus Robotics

The Future of Artificial Intelligence: 2019 Trends and Beyond

Djamila Amimer

- CEO and Founder, Mind Senses Global -

AI is defined as "the science and engineering of making intelligent machines." — John McCarthy, 1956.

It is based on several foundations, such as mathematics, statistics, and neuroscience. In 1956, a program called Logic Theorist was the first artificial intelligence program created. It proved 38 of the first 52 theorems in Chapter 2 of the *Principia Mathematica*.

However, rapid advances in data storage and computer processing power have significantly increased the acceleration of artificial intelligence. Recent examples of advances in artificial intelligence include the development of AlphaGo by DeepMind, which defeated one of the best human players at Go. Nowadays, most of us have encountered AI when using devices and applications such as smartphones, Google Assistant, Amazon Alexa, and chatbots.

AI OPPORTUNITIES

AI market value is expected to reach $60 billion by 2025, with the race between the United States and China being the most prominent. AI is expected to be used in different sectors such as:

— **Natural language processing for voice recognition.**

— **Medical diagnosis and computational biology for cancer detection and drug discovery.**

— **Finance applications such as stock trading and credit scoring.**

— **Image processing and computer vision.**

— **Automotive and manufacturing applications such as autonomous cars and predictive maintenance.**

BCC research forecasts that AI digital assistants will have the highest growth rate over the next five years. Looking over the period 2019-2024, autonomous robots, digital assistants, and neurocomputers will have the highest share of the smart machine market.

Artificial intelligence is expected to make significant impacts on the following segments of the economy: treatment and prevention of diseases; sustainable energy, which includes cleaner energy innovation and a smarter grid; agriculture and food distribution; transportation, including driverless cars and efficient traffic; and robotics for commercial and personal use.

TOP AI TRENDS FOR 2019

The application of AI will continue to expand to different sectors. In 2019, we will see both businesses and governments strive to use AI to improve performance and save costs. More AI platforms and machine learning tools will be developed and deployed to cope with demand.

Specific trends to expect in 2019 are:

1. AI Chatbots and Virtual Agents

The use of AI chatbots and virtual agents will be more prominent in 2019. More companies are expected to customize their AI chatbots and virtual agents. The adoption of these applications by businesses will significantly improve customer experience and branding. The use of AI chatbots and virtual agents will see the most increase in the financial services and retail sectors.

2. Natural Language Generation

2019 will see more data translated into useful texts, reports, and visualization. These types of AI applications will benefit most sectors. Industries such as finance and travel will use them to produce more accessible and effective market reports and travel brochures.

3. Voice and Speech Recognition

Over the last few years, we have seen a rise of voice and speech recognition through the digitalization of call centers and the deployment of devices such as Amazon Echo. This trend is expected to continue to increase. In 2019, we will see more businesses adopting digital voice interfaces to interact with customers.

4. AI and Internet of Things

The combination of AI and Internet of Things (IoT) is powerful. Over the last few years, we have seen the emergence of devices such as digital thermostats and security cameras that can be controlled virtually. This trend is expected to continue to increase with the use of AI virtual assistants to control home appliances and vehicles.

5. AI Hardware

We have seen in the past the emergence of AI hardware and chips from Google, IBM, and Intel. This trend is expected to continue with the use of powerful AI mobile chips becoming standard practice.

AI CHALLENGES

AI is still in its infancy, and its associated risks and issues are not well understood. AI's main issues and challenges are around transparency, bias and accountability.

Lack of AI Transparency

"Black box" type of thinking has been the main criticism of AI. Not all AI and machine learning techniques are interpretable. One specific example is deep learning. With the huge number of layers used to construct the neural network and limitations in modelling cause-and-effect relationships, it is almost impossible to explain how outcomes are derived by deep learning technique.

AI Bias

We have seen several examples of AI bias and failures over the last few years. Examples include the Microsoft bot Tay that started to come up with racist remarks within 24 hours of its interaction with people, Google facial recognition failing to recognize people from different ethnic backgrounds, and most recently Amazon's recruitment tool, which favoured men over women applying for technical roles.

AI Accountability

When it comes to AI products and applications, it is not clear who is accountable. Several parties could have some degree of accountabilities, including data scientists and product producers, as well as owners, users, and regulators.

Digital ethics and privacy are expected to be a prominent theme in 2019. The above issues and challenges need to be addressed if we want to see AI flourish in the future. ~

Dr. Djamila Amimer is the current CEO of Mind Senses Global, a boutique AI management consultancy that aims to reshape business with artificial intelligence.

Robotics Growth in the Year Ahead

Maciej Lisiecki

- Managing Director of Fox Automation, LLC -

What is a robot? The word itself means "forced labor."

It comes from the Czech term *robota* and means "work." It is exactly the same in Polish. On the other hand, Wikipedia describes it as "a machine — especially one programmable by a computer — capable of carrying out a complex series of actions automatically." And that is it.

We have a job to do, and we have a machine equipped with a computer that we can program (force) to do the work. The work has to be done! Our civilized world demands more and more goods, which simply means labor. Simultaneously, society is getting older, and young people are not happy to work overtime.

Today people are looking for a good work-life balance. Such needs require a flexible solution, and that would obviously be robotics. And we have been familiar with robots for decades.

Until now they have done the heavy and the dirty tasks. They have been simple and they have done repetitive work.

Nevertheless, from today's perspective, it is not enough to use the robots for only predictable tasks. The new demands dictate: be ready for the new, adapt, collaborate, customize. Additionally, there are areas where robotics has not been implemented on a large scale — that is, warehousing, hospitals, or deliveries.

The coming year will deliver new technologies in robotics. Robots will address the most fundamental needs, such as adapting to performing tasks and collaborating with people, as well as replacing human hands and vision.

Right now there are two main technologies that determine robotics in this area: collaboration and AI.

Collaboration lets the robot get out of the cage and reinforce human work whereas AI puts a soul in the robot and enables its customization. Robots will have a similar impact on development that electronics once had on automation. They will go even one step further; they will make the work done by robots similar to the work done by humans. The collaborative robot with intelligence will be a perfect solution for an adaptive task where humans "force the labor."

This is the reason why we have seen many new companies being founded — to implement AI in collaborative robots.

The Horizon company has started a project for which a Universal Robot is used and machine learning algorithms are implemented to teach the device how to grip things, just like a small child learns everything from the very beginning. This is a response to one of the key demands for labor in terms of recognize-pick-place. This is what humans do perfectly with their hands and — until now — machines have been unable to.

Every warehouse, every factory has thousands or even millions of pick-and-place movements. Cobots with AI will make it adaptive, and they will slowly replace human hands and enable people to move to more advanced jobs.

Another rapidly growing area is robotics in warehousing.

The coming year will bring fast development of autonomous robots, especially in grocery business. The fact is, they are not really automated. Even more, there is still something that stops people from buying online. There already exist some companies and startups that try to implement a solution for these businesses — that is, adaptive robots cooperating with humans. Robots equipped with AI will learn customers' behavior and adapt to their expectations.

Additionally, they will respond to customers' lack of patience and the fact that the grocery business does not forgive delays. This means there will be a major disruption in the industry. The robots will help to fulfill orders, and AI will perfectly recognize and meet customers' needs.

Being Knapp's partner and a member of the Knapp group, the Horizon company is strongly involved in the implementation of such a solution in the US market.

We are planning to start it on the ground in Poland, too, right after we complete our case study research and we have formed a complete understanding of what drives people to buy online. Automatized and robotized grocery will bring a real revolution in the following year. It will change the world of shopping. Amazon Go, for example, is one of the disruptors.

Thousands of years ago, life was completely different.

The caveman woke up, went into the forest, collected berries, and went back into the cave. The truth is, the environment has changed, but our instinct has not. And this is the area where robotics fits in. Digitalization, automation, robotics, AI, and Fintech will create conditions and rules akin to the ones that controlled the way of getting goods in the dim and distant past.

One more thing we have to bear in mind is quantum computing — a very important future tool for AI development. It will have significant repercussions for robotics, comparable to the ones airplanes and rockets had for traveling technology.

So far, it is still in a very early stage, but still we cannot ignore it because inevitable progress in this field will make real the ideas we have only seen in Hollywood movies.

But it might happen sooner than we expect.

And when it does happen, computing power will become huge as it will enable AI to operate similarly to the human brain and make decisions on the same basis human beings do. ~

Maciej Lisiecki is the Managing Director of Fox Automation LLC and CEO of Horizon-Automation Sp. z o.o. member of KNAPP Group. He was born in 1975 in Poland, and he is an enthusiast of robotics and automation. He is also a graduate of the Wroclaw University of Technology.

The Push Toward Quantum Computing

Brian La Cour

- Head of the Center for Quantum Research at ARL:UT -

The current trend towards greater automation and robotics demand ever-greater computational resources. Sensors are becoming ubiquitous, and information is flowing in from every imaginable source. At the same time, it is starkly apparent to anyone who has purchased a computer in the last few years that they are not getting any faster! Remote server farms from companies like Amazon and Google consume vast and ever-increasing amounts of power. Worldwide, this "tsunami of data" is expected to consume one-fifth of the world's energy use by 2025.[1]

This trend is utterly unsustainable, disastrous for the environment, and completely inadequate to meet our future computational needs. We are heading toward a world in which we are awash with data and unable to process it all.

Clearly, a new paradigm in computing is needed.

Since computers first emerged in the early 20th century, they were based on a computational model that remains largely unchanged to this day. Nowadays, when we think of information, we invariably think of *digital* information.

In the current digital paradigm, information is represented logically in discrete elements referred to as "bits" (0s and 1s) and represented physically as, say, high and low voltages in an electronic circuit. Computers manipulate these physical systems to realize basic logical operations such as AND, OR, and NOT in infinite variety and complexity.

From these simple operations come all modern information technology and the computational tasks to which we have become accustomed. In its many myriad forms, this is the basis of our current Digital Age.

To process information faster and more efficiently, we need to think about information differently. It is important to realize that information is not just digital and discrete; it can be analog and continuous.

The futurist George Dyson was asked recently what computing would look like in the distant future. He replied, "In 150 years most of the important computation will be analog computation..., and the notion of all-digital computation will be a quaint relic."[2]

The idea of analog computers has been around for a while but fell out of fashion in the 1970s with the rise of digital integrated circuits. Digital computers are attractive for their versatility: any task can, in theory, be solved by them, and their precision is limited only by the number of bits you have to work with. Analog computers, by contrast, tended to be custom-built devices used for solving particular problems and were not very extensible.

The cost of generalization was inefficiency: digital computers use brute force to solve problems; analog computers use finesse.

The idea of old-fashioned analog computing is seeing something of a rebirth in the most unlikely of places: the world of quantum physics.

In the quantum world, information can be represented *continuously* in the physical states of atoms and other quantum-scale systems. In this context, one speaks of quantum bits — or "qubits" — representing a continuum of states rather than the discrete, digital values we are familiar with.

A quantum computer can, in theory, represent and process vastly more information than a digital computer. It is distinctly different from a traditional analog computer in that the answer you get upon measurement is, again, digital. The underlying analog state prior to measurement is both unknown and unknowable.

Quantum computing had its origins in the theoretical work of the 1980s and '90s but has experienced an explosion of practical development in the 21st century. Major industry players such as Google, IBM, Intel, and Microsoft are leading the way in technological development. At this time, many in the field are coming to believe that we are reaching an inflection point in the technology and may see real-world applications in the next 10 to 20 years.[3]

Initial demonstrations are already underway, and it is quite possible that an announcement of "quantum supremacy" will be made in the year 2019.

A key challenge for developing quantum computing is scaling the technology in the presence of noise. Quantum systems can be quite fragile, and interactions with the surrounding noisy environment can cause them to "decohere" and lose their efficacy. Isolation can create a large footprint, with many devices requiring sophisticated cryogenic systems or high vacuum infrastructure in which to operate. At the same time, we must interact with them in order to use them!

As the systems become larger, this problem becomes more complex. The benefits of quantum computing rely entirely upon the assumption that qubits are scalable, but whether this is true in practice is as of yet unknown.

A relative dark horse in the race to quantum supremacy is *optically based* quantum computing.

This approach uses photons, so-called particles of lights, to represent and process quantum information. Although long thought to be an unscalable technology, optical quantum computing has rapidly advanced with the appearance of high-efficiency photodetectors and photonic chips capable of shrinking tabletops of optical devices onto a thumbnail-sized silicon chip. The technology is appealing, as it can operate at room temperature and without the cumbersome supporting isolation infrastructure of competing technologies, but it is in its early stages of development and still faces many technological challenges. The University of Bristol has been pioneering this approach for several years and has recently spun off a startup company, PsiQ, that is expected to expand significantly in 2019.[4]

The development of quantum computing has inspired some near-term technologies as well. One concept that my group at the ARL:UT Center for Quantum Research has investigated is the notion of *quantum emulation*.

Using a blend of classical electronics and modern quantum computing, we use analog signals to explicitly represent a quantum state composed of several qubits. Standard analog electronic devices are then used to manipulate these signals, mimicking the behavior of a true quantum computer. Information is represented in the rich frequency content of the signal, much as your voice conveys the expressive meaning of your words. A quantum emulation device, as we have come to call this technology, is capable of doing any operation that a true quantum computer can do, and it does so in the same manner.

Because information is represented in the frequency spectrum of the signal, scalability is limited by bandwidth. The technology is nevertheless appealing for several reasons.

First, it can easily be manufactured on a chip using standard semiconductor lithography techniques and operated at room temperature. This makes it amenable to use as a co-processor in existing fixed and portal devices, like tablets and phones. Second, it does not require many qubits to provide a significant computational advantage. We have estimated that a device operating in the frequency range of 1 MHz to 1 GHz, which is easily achievable, could show a hundredfold increase in speed for solving problems relevant to operations research.[5]

Finally, we predict that such devices, built at a small scale, could become ubiquitous as "personal quantum computers" to be used both for educational purposes and for introducing the broader public to quantum computer programming in much the same way that Apple did for personal computers in the 1970s. Indeed, we have already built prototypes of such devices and plan to use them to introduce high school students to quantum computing starting in the year 2019. ~

Brian La Cour is a Research Scientist who directs the Center for Quantum Research at Applied Research Laboratories, The University of Texas at Austin, where he and his team develop quantum emulation devices and other novel quantum and quantum-inspired technologies.

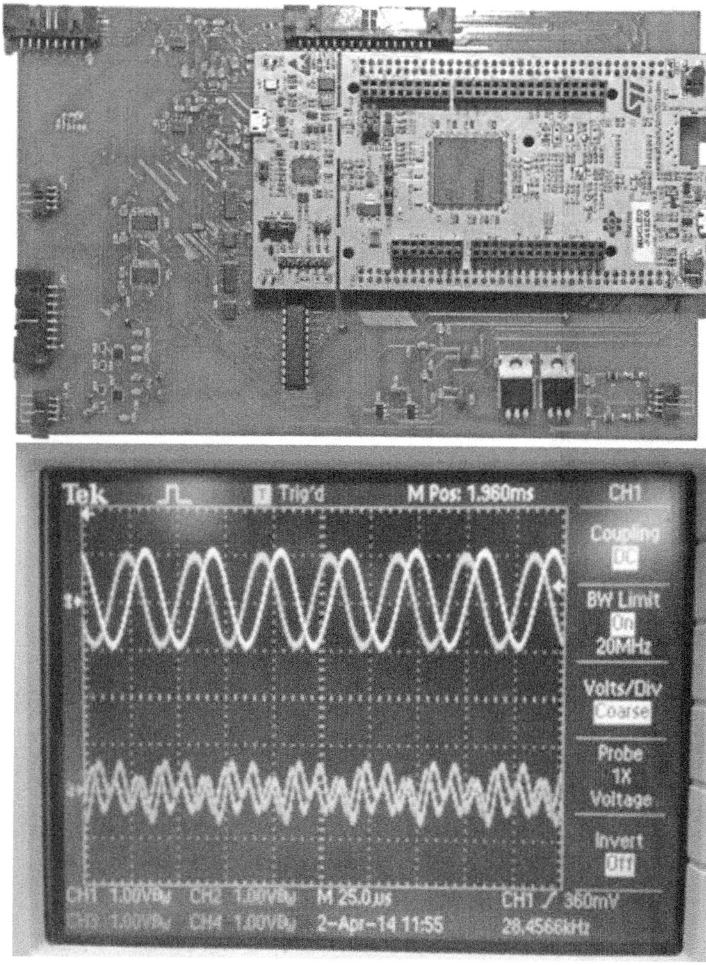

I/O board and signal display for prototype quantum emulation device developed at Applied Research Laboratories, The University of Texas at Austin. (Photo courtesy of ARL:UT)

ENDNOTES

1. https://www.theguardian.com/environment/2017/dec/11/tsunami-of-data-could-consume-fifth-global-electricity-by-2025
2. Scientific American vol. 308, no. 1, p. 36 (January 2013).
3. http://iopscience.iop.org/article/10.1088/1367-2630/aad1ea
4. http://www.psiquantum.com
5. http://dx.doi.org/10.1109/CICC.2017.7993631

Socializing Robots

FI THE FUTURIST INSTITUTE

Evolving the Perception of Robots

Russell Nickerson

- Founder of Collabots and Lab Manager of MassRobotics -

Would you like to know what the robotics community is like from the trenches? Are you sick of the same stories in the news of robots taking over the world and stealing jobs? You probably aren't. In 2019, the same comments will be made about robots that have been for the past century. Committing my career to robotics has led me toward very interesting discoveries about perception. I'm not talking about perception in visual algorithms or sensors but perception of robots as a concept.

I suppose I can largely say that movies are what got me into robotics. As a child I loved *Short Circuit* and *Terminator*. One is a tale about a military robot that gains sentience, and the other is...essentially the same, albeit with a more jovial tone. We are taught robots come from the military. In reality they do as well, because who else has funding for cutting-edge electronics development?

In many fictional tales, at some point the creator loses control of what they create. Kids get old; they move out and make a life for themselves. This common narrative is a springboard to a more compelling reflection on what happens when humanity itself "grows up." This can grow into an intellectually compelling discussion, but often it is lost in the wanton action sequences and broad self-antagonizing of the state of humanity.

"Skynet, man."

Just one pop culture word is used to summarize the status quo. I've attended many trade shows and booths demonstrating robots of varying shapes and sizes. Over the years, I began to see a pattern. I was living my own Groundhog Day. It wasn't everyone, but often the general public would spout something like "robots are going to take over" haphazardly. It is almost exclusively in a tone that suggests jest.

It is almost like people know that robots will take over and it won't matter because humans will prevail anyway.

You know, like the movies.

I've been trying to change this viewpoint. I realize this is a hill that is too steep to climb in my lifetime but still an effort worth moving toward in 2019 and onward. My goal is about as hard as it would be to get everyone to start referring to cats as dogs. I don't wish to overthrow the culture of robotics and everything that goes with it. I rather want to help it evolve into a more realistic place that helps humanity.

A crowd gathers around the Softbank Robotics Pepper robot at an event at the MIT museum.

I started Collabots with the message being "Enabling the success of social robots through collaboration." The core of what I am trying to do is to make people understand that robots are not a human, a slave, a pet, or like anything they have a conceptual place for yet. They are a new social entity. Defining robots is always an important exercise because there aren't many who agree exactly on definition. When I teach children, I ask them if they would consider their dishwasher a robot. Why not? It automatically washes dishes as a task, but it doesn't look like WALL-E. An analogy I explain next for understanding where robots should fit is to imagine you are at Disney.

There is Mickey, which you know is a person in a suit. You play along with everyone and talk and act as if Mickey was that character brought to life. However, if something goes wrong, like the costume head falls off and lands on a child, you realize this is a person representing the work of many people and you can act as contextually is appropriate. The key element of this is the suspension of disbelief with "playing along" contextually.

Right now, the general public is not well informed of what robots do (often the fault of the creators), so they aren't sure how to act toward or around robots. Of course, not everyone should be "forced" to play along with an illusion, but the robot's place and function works better if it is common practice.

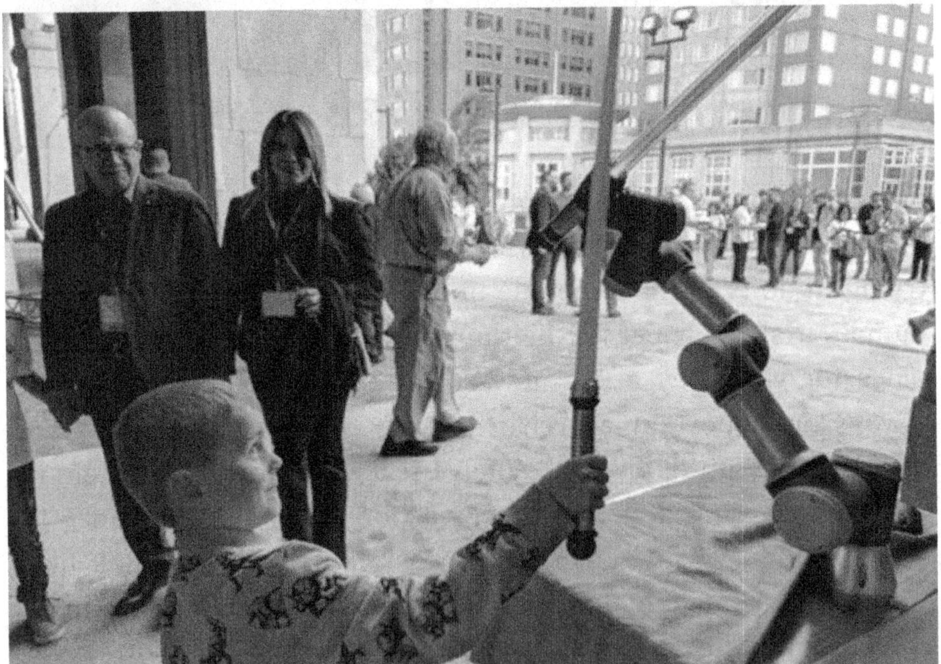

Mock lightsaber battle with a Universal Robotics arm during the Mayoral Conference.

Concert photograph of virtual pop star Hatsune Miku.

Another case, which I would consider almost the "poster" success of the concept I am trying to convey is the virtual idol Hatsune Miku. Miku started out as a synthetic Japanese voice software with a mascot character. Artists, programmers, dancers, and musicians all came together to create different music with the shared persona of Miku. This created amazing content and led to concert appearances via an elaborate projector arrangement, with real musicians as backup on stage. The majority of Miku fans realize she is a character assembled by many pieces of work from a variety of people. Anyone could make their own contribution to the virtual idol scene.

But people may think fans are delusional.

If you understand the context of what it takes to create a character and that real people are behind it, it commands a level of respect akin to appreciating any art.

One of the key aspects of understanding a robot as a character is understanding that many people contributed to its creation and allaying this with respect as you interact. In a way, aren't we alike the sum of a collaboration? We are formed by our experiences and people around us. Robots are formed similarly, either by code, design, or by data accumulated by sensors. It will be a turning point when the day comes that robots can also identify this change in people. In the meantime, expectations have to be managed so that people aren't confused. Robots need to sell for a company to be successful. What happens under ever-present monetary constraints? Over-promises, talking up, and black box features that are nothing more than smoke and mirrors. This might make sense to a seasoned marketing executive, but for those who know the real state of robotics (perpetually not there yet, because *we* as a culture aren't there yet), it is a short sighted and cheap way to establish a product.

Often the nature of business doesn't flow the same way the nature of culture does. I've seen promising and unique products that were ahead of their time perish due to the necessity of business. We've all heard of crowd-sourced projects that fall flat on their promises due to inexperience and unrealistic planning.

Unfortunately, in robotics, engineering and sales don't combine to make the best cocktail.

Sales demands frequent demos for proof to sell; engineering wants to work on the real problems that take time. At some point, these come to a head in a traditional company, and the product created is the product pitched. Too often in robotics, a product created falls far short of the expectation and hype.

Robotics is afflicted with this problem that it is easy to impress people that don't truly understand your product.

It is insurmountably harder to meet the expectation of the public. And so, the loop continues: develop a demo that can easily impress, marketing promises the grail, and the R&D team makes the best with what they have. If the ultimate robot could ever be created, it wouldn't be within a consumer price point. Unfortunately, even the best robots of today aren't enough to satiate the consumer whose expecting Rosie the robot from the Jetsons for under $1,000. (How is she even still relevant?)

For those worried about a robot uprising, humans always will have at least two qualities that robots will not. Versatility and creativity. Being versatile means that physically and mentally you can approach different tasks on a rapid basis. While not amazingly glamorous, fast food workers can cook, clean, manage, calculate, and handle even the most unexpected situations. Even robots programmed to do many tasks need repositioning and reprogramming to work on a new task.

Being creative is coming up with new ideas or solutions for situations we encounter in our environment.

We all see the world a bit differently and can learn and benefit from these different perspectives. The idea of machines being perfect or immortal is also a fallacy, because every system needs to be serviced and maintained. As humans we aren't perfect, and a robot is even more limited and narrow skilled than us.

Until people understand robots better, as in where they fit in society and how to frame them, we will continue to be duped. What is it that we want in a robot? Fiction raised the bar too high, and it is time to lower it. In 2019, robot toys will come out, artificial intelligence will answer neutrally to leading questions, and voice-controlled internet of things buzzword beasts will reign. Advancements will be made, sure, but until robots are perceived as a new, unique entity, we are stuck with expectations that may be forever too great. ~

Russell "RoboRuss" Nickerson is the founder of Collabots, where he does robotics consulting, training, and speaking. Russell is also Lab Manager at MassRobotics, a non-profit robotics coworking space located in Boston's Seaport. He has previously worked at Softbank Robotics/Aldebaran on the Pepper and NAO robots.

Bridging Industry and STEAM Education

Jason Gouw

- Founder of Cubit -

As I grab the door handle to my car, the locks automatically open. I sling myself into the comfortable leather seat as it perfectly cradles the contour of my back. I depress the brake pedal and push the "Start" button. The engine roars to life, and a screen on the dash illuminates — displaying information about my car, the air temperature, my location, traffic conditions, and much more. I now have over 400 horsepower at the disposal of my right foot, ready to tackle the quickest of roads and freeway on-ramps.

The purpose of technology is to make human life "better" through an improvement of convenience, speed, efficiency, or capability.

When I touched the door handle of my car, capacitive sensors embedded within the door handle triggered a computer module in the car to scan for a valid keyfob by transmitting an RF "ping."

The keyfob, in my pocket, received the "ping" and transmitted an authentication code back to the computer module, where it was validated, triggering the opening of the door locks. This all happens within the blink of an eye.

When I pushed the "Start" button while depressing the brake pedal, a signal was sent to another computer module — the engine control unit — where it once again authenticated my keyfob before automatically starting the engine. To keep the engine running, the engine computer has to communicate with several dozen sensors to analyze conditions such as temperature, air pressure, and fuel quality. It performs millions of calculations every second to determine optimal parameters for fuel timing, valve timing, spark advance, fuel metering, and exhaust oxygen levels to name a few.

For me, all I had to do was push a button once and step on the accelerator pedal if I wanted to increase or decrease engine power. Although engine control systems are yesteryear's innovation, it is a great example of how this technology has contributed to the convenience, speed, and efficiency of human transportation. Its integration into the human culture is incredibly seamless and subtle, despite doing so much in the background.

It has become ubiquitous in our everyday lives.

Fast forward to modern-day technology — we have a conglomeration of advancements in hardware and software that allow us to do remarkable and amazing things in a viable and economical way.

Cars are capable of driving themselves and avoiding hazards. Robots are used in warehouses to lift, sort, and transport merchandise autonomously. Homes are becoming automated with integrated systems that learn human patterns in order to optimize heating, lighting, and security with hardly any human input. The same formula applies here - improve human convenience, speed, and efficiency.

So how do companies come up with these ideas?

Something often missed in the world of engineers and developers is the understanding that bigger isn't necessarily better. Faster does not always equate to increased productivity. More powerful motors and faster processors do not linearly translate to a technology that improves our lives. Just because we can create a sweeter sugar doesn't mean it's going to make a better cake.

In fact, technology has equal potential to improve our lives or make it complicated. It takes thoughtful consideration in order to properly integrate these technologies in a way that produces a truly impactful end product.

As I walk through rows of robotics companies displaying their latest and greatest, I am stunned by the larger-than-life creations, but I have a harder time understanding how much of a positive impact these innovations are truly going to make. I've lost count of the many different robotic mobile platforms out there.

Some have four wheels, some eight, some treaded, some have a little bit of this, a little bit of that, and almost all are controlled via a hand-held remote control. To be honest, my first thoughts were of the radio-controlled cars I played with as a kid, only this time, the components are upgraded with more powerful and precise motors in a much more robust and configurable infrastructure.

But what is the application and value add in which these platforms would be used to improve human life?

Are we just creating a sweeter sugar here, or is the final cake actually going to taste better?

The smartphone is perhaps one of the most impactful pieces of technology developed in the last decade. It is used by just about everyone on earth, including developing countries. If we break down the technology that goes into a smartphone, we find that it's broken down into many different categories.

Smartphones have a camera, a screen, haptic feedback, biometric sensors, sleek ergonomic design, speakers, an operating system, music player, Wi-Fi, Bluetooth, flashlight, cell phone, and much more. There is nothing truly special about each of these pieces of technologies, but when you integrate them in a single, user-friendly package that is highly configurable, it becomes something amazing. The secret in today's advancements lies in the integration of technologies, not just a single piece of technological advancement.

In the education system, the integration of technologies is the driving force behind STEM and STEAM education.

STEM is an acronym for Science Technology Engineering and Math.

STEAM is a more recent acronym that adds "Art" to the acronym.

During the formation of my STEAM education company, I did a lot of research and met with numerous teachers and directors from the education sector. I never heard of a clear, definitive answer to the driving force behind STEAM.

Things I hear often relate to "what it is" and "how we implement it in our schools," but the question I was looking to answer was why.

I believe that STEAM education is directly connected to our current industry — also known as the fourth industrial revolution. In the past, we had the first industrial revolution of steam power used to power ships and locomotives. Then we had the second industrial revolution with the innovation of electricity. The third revolution was the innovation of computers and how they helped us automate and accelerate many processes. The fourth industrial revolution, however, does not highlight the birth of a single piece of innovation like revolutions past; instead, it is defined by the integration of existing technologies.

It is this fourth industrial revolution (4IR) that is the fusing of the physical, digital, and biological worlds.

The US education system is a little slow to catch up with industry progression. K-12 education realizes the need for learning technology to better prepare our youth for a successful career but also unsure of how to really go about that. What we find in the education market are countless "STEM products" that are typically coding- or robotics-based, yet advertised as a STEM or STEAM platform.

Though coding, programming, and robotics are trendy buzzwords indeed, they are not STEM or STEAM — only a small facet of the overarching concept. If you look at software developers of today and the future, they are no longer creating code that operates strictly within a computer. Much like the computers in our cars, the developers of today create code to control cars, rockets, and physical robots that have to interact with the physical environment as well as biological users — humans.

These programmers must also understand the physics and dynamics of their application. In order to fully comprehend STEAM, we must begin teaching kids not only the fundamental concepts in each subject area but more importantly how they can be tied together to achieve even greater things — things of 4IR.

Another point of confusion within the K-12 education sector is the understanding of the "Art" component of STEAM.

If you ask around, many will likely define it as a way to express your creativity. But if we examine the relationship between 4IR and STEAM, the Art component is very much a key component in fusing the physical, digital, and biological world. It is not a random smattering of colors, nor is it necessarily decorative. Instead, the Art in STEAM refers to creative design that enhances the user experience. Design that helps integrate the human-machine connection. Design that fuses the biological component (humans) with the digital and physical technologies.

Take a look at some of the best products today — from smartphones to laptops to cars. Chances are, there was a fair bit of thought that went into creating an elegant design that seamlessly meshes the human-machine interaction.

In 2014, I started a STEAM education company called Cubit, specifically for the K-12 education market.

With my background in robotics and mechatronics, I saw an opportunity to bridge industry and STEAM education. Our approach to STEAM education is much different than what I've seen in so many other "STEM" products. Instead of using ready-made click-together modules for building different STEAM projects, we emphasize the use of raw materials. Our projects are made out of cardboard, construction paper, Pringles cans, empty water bottles, 3D-printed pieces, and the like.

Through this approach, we are able to challenge students to think deeply in order to figure out how to build with these raw materials.

In contrast, click-together blocks such as Legos are easy and fast to build with, but at the end of a lesson, students typically learn to follow assembly instructions instead of thinking creatively on how to build a water turbidity monitor out of an empty Pringles can and water bottle.

With our approach, the learning doesn't end there. After students build their projects, such as this water turbidity monitor, they are given a design challenge. In this case, students are given an assortment of materials to design and create the best water filter. They can use materials such as coffee filters, paper towels, gravel, sand, etc., to create a filter. Some dirt is introduced into water and analyzed in the turbidity monitor, where students take a baseline measurement. Students then run the sample water through different filters of their own design and observe the change in turbidity.

What we've done in this single curriculum project is to teach kids about water pollution in our environment, engineering a test instrument, creating the programming for the test instrument, designing a solution around the problem, and using mathematical analysis to interpret data in order to make improvements and iterations to their design.

This is what I believe to be true STEAM education — teaching kids how to integrate the various skills within STEAM in order to solve real-world problems. May we properly educate and invest in the future of our world through our youth. ~

Jason Gouw and his brother grew up with a father that engaged them with hands-on projects made from scraps. Jason grew up to teach himself electronics and robotics and started his own design firm at 18, where he prototyped custom electronics for various companies such as GE and Hallmark. In 2014, he started Cubit, a STEAM education company for K-12 education. He hopes to bridge what he's learned in the industry and bring it into the way we educate our kids about the future of technology.

Every Chatbot is Now a Voicebot

Lorenzo Carver

- Founder of SpeadBot and CoinSwipe -

Have you ever seen a child that can hardly speak successfully interact with an iPhone?

It's kind of amazing that a visual user interface could be so intuitive that it could communicate with a person that can't read.

As impressive as advances in GUI design have become, seeing smart speakers such as Amazon's Alexa in action make it clear that a massive shift in how we interact with data, applications, and other human beings is underway. In 2019, this shift will accelerate as text based chatbots used for e-commerce, eHealth, eLearning and Fintech.

Conversation robots, or chatbots, are computer programs that conduct a conversation or perform a task via text.

Studies show that people who interact with a conversational robot tend to expect the robot to do human things as well as be able to say human things. In most cases, chatbots fail to deliver on this expectation. Despite these failures, market research firms have predicted year-over-year growth of 25% higher through 2024.

The emergence of a "killer" chatbot would of course change this growth trajectory for the better.

And that didn't happen in 2018. Instead, something far more powerfully beneficial to every chatbot developer occurred.

Chatbot delivery platforms with highly accurate natural language processing (NLP) and powerful natural language understanding (NLU) capabilities have turned tens of thousands of text-only chatbots only fraction of target users would enjoy into potentially millions of voicebots over 50% of mobile users are already used to interacting with.

The repercussions of this shift are particularly profound in the United States, where text-based commerce has never come close to what it is in markets such as Asia. WeChat, Tencent's 1-billion-user chat app, has over 500,000 chatbots built on it. With an integrated wallet that users in China used to make trillions of US dollars' worth of payments, it's clear that the trust of text-based commerce associated with WeChat opens the door for higher adoption of new chatbots on the platform.

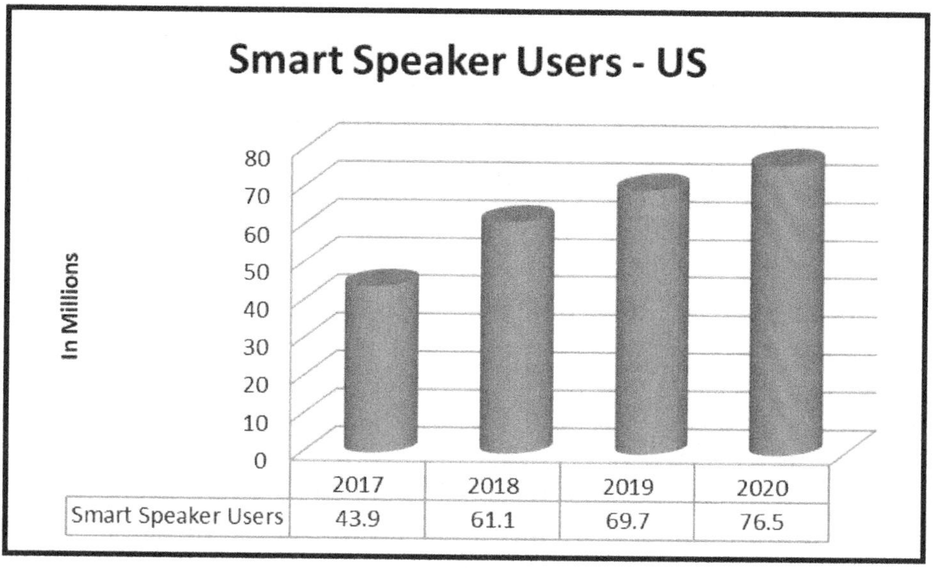

Source: eMarketer April 2018

In the United States, Facebook's Messenger platform, along with its $19 billion acquisition of WhatsApp, was expected to follow a similar path.

But that didn't happen.

Although Facebook was very successful in getting developers to use its platform to build more than 300,000 chatbots today, the levels of user engagements and the depth of use cases are not comparable to what WeChat was achieving years ago. While some will point to shifting levels of brand trust being a factor for Facebook, the inability of Facebook Messenger bots to improve the user experience by some order of magnitude compared to alternatives is clearly a key factor.

If you take everything Facebook had hoped to accomplish with its Messenger bot platform in the first year and multiply that by 10, you will be begin to see the potential of Alexa's voicebots (branded as Alexa "Skills") market this coming year. Amazon already has payment information on over 300 million active users, so the connection between the credit cards, their physical wallets, and the payment information in their Amazon accounts is already established.

With limited development effort, Facebook accounts can be "linked" to Alexa Skills. There are two natural extensions of this. The first is to generate Alexa Skills for most business and e-commerce websites. This is already happening, and it started with sites that had already made efforts to add more text-chat functionality.

For those that believe in the power of distributed ledger technology, such as blockchain, the second is to incorporate web.js, or comparable, technology to turn those voicebots into blockchain-enabled decentralized applications where value can be exchanged directly between the site visitor and the content, product, or service provider.

Here's an example of how combining Facebook Messenger for identity and certain chatbot functions with Amazon Alexa for voice can improve the lives of users and business owners with little, or in some cases no, developer time expended.

"Alexa, complete my taxes."

Upon hearing this, Alexa could list a number of tax preparation skills available, including those from large service providers and those from local shops. However, that would be a cumbersome experience for users to sift through and then decide which tax skill, or service, was best suited for their budget, level of income, and level of complexity.

A bot that asks for permission to use their purchase history on Alexa to get a match might use "Complete" as a keyword trigger and respond with something like, "Hello George, this is TaxVoiceBot, I've found ten sources to complete your taxes for between $2 and $300. Would you like me to recommend the top three best matches for your income and budget?"

If George says "yes," TaxVoiceBot responds, "Great — I'll need to look at your purchase history and zip code to estimate an appropriate budget. Is that OK?" Assuming George proceeds, TaxVoiceBot can automatically place tokens from one of George's wallets into a smart contract that will be released to the tax preparer matched once TaxVoiceBot sees that the return is ready for review and filing. George never types anything in this process and may not even need to read anything either.

Over the past half year, we've built three systems that are similar technically to what is described above (although not for a tax return use case).

In each case, the Alexa Voice piece was the *easiest* part to implement.

Based on that limited firsthand experience, we believe other chatbot developers will follow a similar path, so that very quickly (perhaps this year) the number Alexa Skills will exceed the number of Facebook Messenger bots because every Messenger bot will be converted into an Alexa Skill. ~

Lorenzo Carver is an Inventor and Entrepreneur currently developing conversational robots. He is working to combine chatbots and blockchain technology to make complicated tasks easier for consumers.

Humans and Cobots: Who's Collaborating with Whom?

Keith Blanchet

- Kinova Head of Robotics Innovation and Global Sales -

We expect 2019 to be a year of significant change in the robotics industry, in particular when it comes to the adoption of personal, or service, robots.

In our industry, the term *cobot* is used to refer to "collaborative robots," devices that inherently involve human-robot interaction.

But I question if the term *cobot* is actually correct.

When a human and a robot are working side by side on a project together, who's collaborating with whom? Is the robot working alongside the human, or is the human working alongside the robot? Who's master and servant?

It's a fundamental misperception in the service robotics industry.

In true human-robot interactions (HRI), the robot should always be the one listening to and learning from the human — not the other way around. When the design thinking is more about automation, it's intent is to replace humans, not amplify their capabilities. It puts the robot at the center of the equation.

Robots should be amplifiers, not replacements.

Cobots or Assistants?

What's in a name? Well, if you think about it, the cobot collaboration equation is a microcosm for society's interaction with robotic technologies. The mentality of fear around a potential "robotic takeover" will only be actualized if we build robots that work for us, rather than with us. It's a big — yet often overlooked — difference.

A far more accurate name for what is widely perceived as a cobot would be *assistant robot*, if the intention is truly embracing and not replacing human capabilities.

Why do the specifics of how we interact with robots matter?

They matter because society can't afford to make itself obsolete.

A lot of industries are moving toward partial, if not complete, automation; we're setting ourselves up for economic crises as more and more jobs are replaced, not to mention the impact on mental health and self-worth, as well as the risks associated with rising crime rates as unemployment rates increase.

Robotics manufacturers who aren't convinced of the risks of unemployment are going to bump up against another problem: distrust of AI. With societal perspectives of artificial intelligence being...well, more than a little wary, acceptance of robots from an automation point of view is going to meet understandable resistance.

A perfect example is the increasing accessibility of robotics in healthcare. Can we trust robots and artificial intelligence when it comes to making life-or-death decisions? The job of the robot should be to empower health professionals so they can provide better care to more patients, not to have the foresight to evaluate moral implications or make healthcare decisions.

Amplification vs. Automation

Replacing people with robots is a direct line to social unrest, leading to plummeting economic and societal well-being. Robots are important and powerful tools, but they need to be implemented with a watchful eye.

So where do we go from here? How can we convince robotics manufacturers to take the same approach?

The most important step in getting robotics manufacturers on board with human-centric robotics design is finding a way to make them recognize that robots should be amplifiers, not replacements.

What we have, at the moment, are robotics manufacturers operating from a perspective of human adaptation to robots, not the other way around. A robot is implemented in order to automate a task, meaning the human does what it can to collaborate with the former and help it meet its objective.

A robot implemented from the perspective of an amplifier, however, is one that adapts to human beings instead. It needs to be able to listen to a human in a way that matters, and gather the necessary information to aid humans, not command and direct us.

Who should care about this?

Everyone.

Here's why:

If there's one thing people who know they're facing surgery want to feel, it's at ease. And while it would be impossible for patients to go into a procedure with 100% unwavering certainty, it's at least helpful to know that conditions are as optimal as possible.

Aside from just precise cuts, robot technology allows surgeons to have more control over their environment. Whether it's to ensure a patient doesn't lose a lot of blood, is able to recover in less pain, or is able to minimize a hospital stay, surgeries involving robot assistance are proving to create the right circumstances for people to feel at ease.

Or what about the use of robots to enhance the safety of humans in situations requiring the manipulation of hazardous materials (e.g. nuclear, biological, or chemical) and bomb disposals (like improvised explosive devices or IEDs)?

Humans should never be exposed to the dangers of explosive devices and harmful chemicals, but the truth is that handling them requires human wherewithal.

It's imperative that skilled professionals remain in full control of these high-risk situations to make the right call. Fortunately, they can now command robotic devices to perform the defusal and disposal of such materials and explosives from a safe distance.

Conclusion

As human beings, we're always striving for something else; we always want to learn and experience new things.

The rapid pace of innovation makes it inevitable that we'll develop increasingly complex robots. When developed thoughtfully, dare I say ethically, with human needs at the core, they'll enable and empower us to achieve more.

In the next year, we should be working toward making robots more accessible and accepted, and that comes with an understanding and trust in technology. That's what true human-robot interaction is — man and machine working together for the greater good. ~

Keith Blanchet is VP of Global Sales and Head of Kinova's Innovation Robotics division. With a passion for leading-edge technology, he helps Kinova develop offerings in IoT, AI, Big Data, and the fusion of a number of sensing technologies.

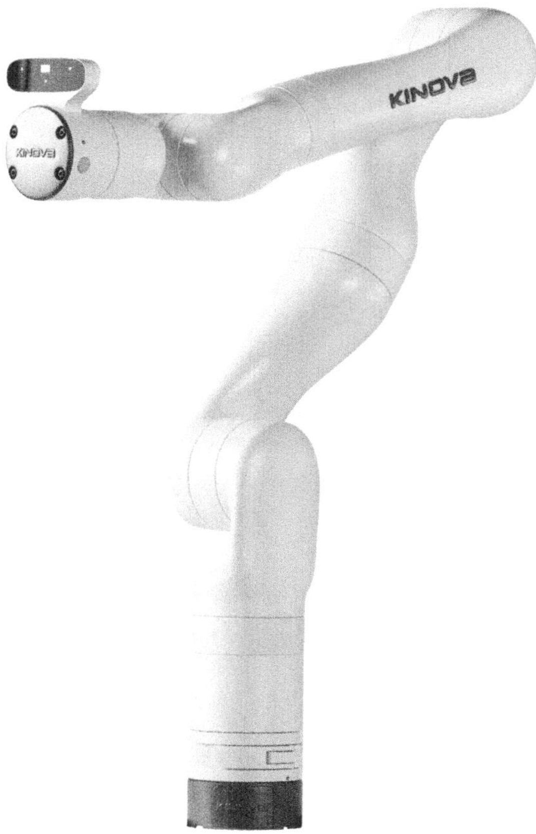

Images provided by Kinova with permission.

The Need to Overcome Interoperability and Labor Replacement Fears

Daniel Theobald

- Co-Founder of Vecna Robotics -

There is much hype and debate in the media these days pitting robotics and automation against jobs.

When someone worries about losing their job, it can cut to the core of their confidence and positive outlook in life. That is part of what makes this such a sensitive and complex issue. The risk is that if we aren't thoughtful about how we approach this issue, we risk doing more harm than good to the very people we are trying to protect. For example, some have suggested taxing robots to pay for certain social programs.

Unfortunately, the idea of taxing robots is almost certainly unimplementable from a technology point of view. It is also, quite frankly, a bad idea.

The short version of human history is simple: **Humans create technology, technology creates prosperity, prosperity is shared, and human society thrives and creates more technology.**

This is the "prosperity chain." Having more means I am likely to share more. Having more also allows me to spend a greater percentage of my time on things other than meeting my basic needs. Note that it is not necessary for me to care about other people for this to work, although I hope to show that if we do care, we will all do far better.

Let's consider some examples.

Recent research has shown that Neanderthals and Homo Sapiens may have co-existed in Europe for over 5,000 years. While most of us have a small amount of Neanderthal DNA, the Neanderthal as a distinct species no longer exists.

Why did Homo Sapiens thrive and Neanderthals disappear?

No one really knows, but it appears that **Homo Sapiens had a unique ability to innovate and create technology that allowed them to prosper.** While this is speculative, evidence is accumulating to show that Homo Sapiens could do a better job taking care of their people over long periods of time.

Early Homo Sapien technology included needles for sewing, improved arrowheads for hunting, and complex communication making trade possible.

Sewing, for example, allowed them to adapt to a wider range of climates and better clothe their people. Clothing is a technology that provides more benefit to the weak and frail than it does to the strong.

Arrowheads allowed for more effective hunting. Bringing home more food meant that more people were fed than just those who could kill their own.

Communication not only allowed for trade and a more stable society but also allowed for the needs of underserved parts of the population to be better understood and met. Technology created prosperity, which was used to benefit the broader population.

And in the history of civilizations, we have seen time and time again that those cultures and nations who embraced change coming from new technologies thrived and prospered, whereas those who attempted to ignore it eventually had to accept it, at times with enormous hardships involved. In today's interconnected world, we do not even have the space for isolationism anymore.

Now, let's take a closer look at the technology that brought us this sudden change: the internet.

There can be no argument that the internet has created prosperity and that there are numerous examples of where that prosperity has been used to attempt to better the lives of everyone on this planet.

Yes, the consolidation to online retail has created challenges for traditional retailers and brings into question what types of jobs may be available in the future...a trend that is likely to accelerate. Yet it is clear that in part because of the internet, the human race, on the whole, has unprecedented levels of freedom, education, equality, basic human rights, and access to capital. The trend here is that the disruption caused by short-term change is the cost we pay for the long-term gains in overall prosperity for the human race.

Now, let's say for the sake of argument that we all agree that prosperity creates the possibility of caring for more people. This should be clear: if I don't have many resources, my ability to help others shrinks.

When we think about people with more resources, it is easy to focus on the widening gap between rich and poor in part caused by greed and lack of empathy for those with less, but I believe these people are the minority (a.k.a. greedy bastards).

The key to balancing the equation is our values. What you do with your prosperity will be determined by your values. On the balance, over the long term, humans tend to have values that help the human race. There are endless counterexamples, but the bottom line is that the human race continues to advance and flourish. This would be difficult in a scenario where people were unable or unwilling to help others.

Technology -> Prosperity -> Sharing -> Stable Society -> Technology.

The prosperity chain acts over long periods of time, while the technological disruption of how humans work is generally short-lived. Societies that fail to prosper collapse. Societies that fail to embrace technology put their long-term prosperity at risk. So assuming taxing robots was even possible to do, why would we do it?

We tax cigarettes when we want people to stop using them, which seems reasonable, but trying to deter the use of prosperity-creating technology by taxing it is simply not good policy.

In the next 20 years, we will see acceleration in productivity growth caused by automation that will also result in significant changes in the global labor market.

There are predictions that half of all human job activities could be automated, yet only 5% of all jobs can be fully automated. About 400 million jobs globally could be replaced by automation and AI by 2030, but predictable investments in infrastructure, energy, and technological development and deployment will offset these losses. And we don't even know yet what 10% of the new jobs will look like — just like the tens of millions of jobs that have been created since the inception of the internet. On the whole, history has shown that technology is a job creator.

But here is the thing: will humanity need everyone "working" to meet our basic needs of food, clean water, clothing, and shelter? Not by a long shot. That was humanity of days past. It only takes a small subset of society to create those things. The rest of us are involved in what was in ancient times generally considered "luxury" industries, like education, science, medicine, travel, art, media, etc. Many don't consider these to be luxuries today, but for most of human history, these pursuits were only available to the elite. Prosperity has made them available to a larger circle of humanity, and that prosperity was the result of technology.

We have basically three options:

1. Overreact like the luddites of old and sabotage the very technology that will allow us to continue to prosper. This will damage our economy and result in long-term rather than short-term job loss.
2. We can continue to improve our technology but forget that we have a fundamental responsibility to care for others. This risks widening the gap between the rich and the poor, eventually leading to increased social unrest or revolution.

3. We can use the increased efficiency of technology and the prosperity it produces to make the world a better place for everyone. Overall, this is what our species has been doing for millennia, albeit with lots of room for improvement. I vote for this option and suggest that we can do this better than ever before.

My hope is that we can use our technology-created prosperity to evolve better socioeconomic models over time that allow the debate to turn from "will everyone have a job?" to "does everyone have the opportunity to do something that they enjoy, and how can we align that with what we need to accomplish as a species?" Along the way, let's make sure to be aware of and take care of those that need a little extra help.

So to summarize: Robots and automation can create unprecedented levels of prosperity. The prosperity chain creates a virtuous feedback loop that allows for more prosperity and creates the possibility to take care of more people than ever before. Ultimately, what we do with that prosperity will define who we are. Let's do good. ~

Daniel Theobald co-founded Vecna in 1998 with the mission of empowering humanity through transformative technologies. Daniel is also Co-Founder and President of MassRobotics, an independent, nonprofit, industry consortium dedicated to fostering the global robotics community.

Robots in Industry

FI THE FUTURIST INSTITUTE

The Rise of Smart Industry:
How Artificial Intelligence Will Power a New Wave of Innovation

Jeff Burnstein & Robert Huschka

- The Association for Advancing Automation -

The motion appears human. Even if the hand is not.

The robotic fingers deftly turn a colorful cube that looks a bit like a children's toy. There's a green "N" on one face. A blue "A" on another. The robot hand is instructed to show a red "O" outward. It quickly obeys, balancing the cube in its palm as its fingers slide the toy around.

No one told the robot how to do this. Dactyl — the software that drives the hand — taught itself.

OpenAI — a nonprofit think tank started by Elon Musk – created a virtual simulation where Dactyl could play with objects and learn from its mistakes. It took Dactyl roughly 100 years of simulated time (50 hours in the real world) to hone its skills.

And during that time, researchers kept messing with the simulation, by making the digital objects heavier or lighter or by changing their shapes or colors. They even altered "gravity" itself.

When it was time to move out of the virtual world, the randomness of the simulations made Dactyl much more effective at handling conditions it would face in real environments.

There have been many robot hands. But most are programmed to move a specific type of object through a specific set of motions. What makes Dactyl interesting is it can generalize. When faced with an entirely new object, it adapts, with humanlike behaviors — using its fingers and its vision-camera eyes — to feel its way through a task.

Dactyl and many comparable efforts represent the next revolution in robotics and automation — machines and processes that can sense, understand, act, and learn.

This is not an entirely new revolution. Versions of machine learning algorithms have been with us for decades. AI is already with us on a daily basis. In our smartphones, in accounting programs, in collaborative robots, and in the smart assistants on our kitchen counter.

But in 2019, we expect the speed of this innovation to intensify.

Don't Let Our Fears Stand in the Way

Certainly, there are lots of reasons to debate the impact that AI will have on society. The conversation we are having now on AI ethics, job retraining programs, and autonomous weapons are important and necessary. As with any new technology, we must learn how to maximize its usefulness to humanity while mitigating the disruption in people's lives. No one is saying this will be easy with artificial intelligence.

But this year, can we please stop talking about the Terminator?

Too much of the conversation on AI is rooted in fear. While there are valid concerns, we frequently lose track of the tremendous opportunities that await us. Artificial intelligence has the potential to unlock amazing benefits for all of us.

Here are some examples:

MAKING MEDICINE BETTER

AI researchers are using tools that can help tell the difference between two types of lung cancer or detect depression in your voice. AI-powered robots may help care for you when you're old or disabled. Today, one in eight kidney transplants (from a living donor) is matched by an AI-powered online organ exchange.

FIGHTING HUNGER

In October of 2018, a coalition — which includes the World Bank, the United Nations, and the Red Cross, along with Microsoft, Google, and Amazon — announced an AI tool that will predict where the next extreme famine will occur. Also, artificial intelligence will help farmers to decide what to plant and when — while limiting the use of pesticides — to boost food production.

MAKING US SAFER

In India, artificial intelligence — combined with high-tech sensors — is being used to monitor the structural integrity of bridges. Credit card companies and banks are using finely tuned algorithms to prevent fraud. And new security systems can detect crime patterns, triangulate the locations of gunshots or help find missing children.

SAVING THE PLANET

Artificial intelligence will give us incredibly accurate models on weather and climate change. IBM's Green Horizon project is letting global planners "test" pollution-reducing strategies to see what will make a difference. Machine learning is helping companies control their energy usage and reduce carbon emissions.

These are revolutionary applications that have the potential to make our businesses and our world better — if we don't let our fears stand in our way.

The Dawn of the Truly Smart Industry

At the Association for Advancing Automation, we have a front-row seat to the emergence of smart industry — a coming together of disciplines ranging from collaborative robotics to machine vision to supply chain optimization and advanced analytics. Artificial intelligence — in many shapes and forms — will be the stitching that weaves together this new age of industry.

This revolution goes by many names: Industry 4.0, The Factory of the Future, Smart Manufacturing, the Industrial Internet of Things. No matter the label you choose, this disruption will touch every aspect of the manufacturing ecosystem, from product design to manufacturing processes to the delivery at a customer's door.

We are already seeing the impact in automation:

PREDICTIVE MAINTENANCE

Machine learning algorithms — fed by an unprecedented amount of sensor data — are taking the guesswork out of plant operations by identifying problems and prescribing solutions before an incident leads to costly downtime.

PRECISE VISION INSPECTIONS

Technologies like lidar, laser, 3D, or infrared, machine vision systems — backed by machine learning — are leading to unprecedented manufacturing quality. These systems can spot defects invisible to the human eye at amazing speeds.

A REVOLUTION IN ROBOT GRIPPING

Robots are learning to lift objects of vastly different shapes, textures, and weights. Combined with vision advancements and AI, robots are deciding the best ways to pick up a bag of potatoes or a coffee cup — without dumping the contents. Soft mechanical hands are feeling their way around before grasping unfamiliar objects.

COLLABORATIVE ROBOTS THAT LEARN

AI-powered collaborative robots — deployed alongside their human co-workers — can sense and react to their surroundings. They can spot nearby humans and slow down or stop to keep people safe. Some robots can learn new skills by mimicking the actions of human trainers - with no coding required. If robots fail at a task, they'll be able to try again, learn from their mistakes, and share that knowledge with their fellow machines.

AN ERA OF AUTONOMOUS MOBILITY

Self-driving cars are on the horizon, but we already live in a world where mobile robots — zipping around warehouses — provide the backbone of e-commerce empires. These robots can shuttle parts around factories, smartly charting their own path while dodging obstacles and human co-workers. Some are already churning in our fields, helping farmers seed crops and kill weeds. And others are headed deep into our mines, saving humans from another dangerous job.

A SUPPLY CHAIN THAT EVOLVES

Cloud-based tools can analyze giant chunks of data from your supply chain and help optimize your procedures. They can ensure you don't understock — or overstock — your warehouse. There's software that can pick the best routes for your trucks to minimize fuel use or can warn you if there's about to be a crunch in air freight availability. Data from your many suppliers can be used to predict a parts shortfall before it shuts down your production. And these algorithms are already predicting consumer demand before a product is even manufactured.

We have only just begun to scratch the surface of what is possible. There will be applications that we have not yet even considered that will shake the industrial landscape in the years ahead.

The Challenge: The Ambition-Execution Gap

Many CEOs already understand they are in a race. According to a study by the Boston Consulting Group and MIT Sloan School of Management, 85% of executives believe AI solutions will give their business a competitive advantage. However, just 39% of companies have started down the road. This has been called the ambition-execution gap.

Companies know they need to implement AI strategies but are frequently at a loss on where to begin. Or they get stuck in a pilot purgatory, where AI initiatives never reach a scale where they can help the overall enterprise. Even when on the right track, they lack the talent in the pipeline to support their efforts.

But the companies that navigate these uncharted waters by leveraging their data and investing in the necessary workforce — and that are brave and nimble enough to evolve — will gain unprecedented advantages. These companies will blend human know-how with AI insights and precision.

Understanding the challenges ahead, the Association for Advancing Automation is launching a major initiative in 2019 to help companies navigate this journey. Microsoft, Intel, GE, Amazon, Siemens, Google's X, and others are assisting the association in an effort to help business leaders understand the evolving AI opportunities and discover valuable use cases that will drive their industries forward.

The future is arriving fast — and it will be automated. ~

Jeff Burnstein is the President of the Association for Advancing Automation (A3), an umbrella trade group representing more than 1,200 global companies focused on robotics, machine vision, motors, motion control, and advanced manufacturing technologies. **Robert Huschka** is A3's Director of Education Strategies. For the latest details about A3 activities, visit www.a3automate.org.

Transportation Automation and the Future of Multiplicity, Not Singularity

Craig Fuller

- CEO of Freightwaves -

We tend to hold the past on a pedestal and demonize the future. If you interviewed ten experts on robots right now, you'd get ten different definitions.

However you define these machines, artificial intelligence (AI) already abounds. Our smartphones might as well be a second brain. The grand sum total of human knowledge is held in our hands — or trembles at our fingertips — at any given second. Why do we not call our Garmin tracking devices robots? Why not our iWatches? What's the difference between what they do and what a robot does?

Robots perform tasks. Robots do what we program them to do. We probably wish they'd do a lot more than what we program them to do, but usually it's a lot less.

It's widely known (at least among science fiction geeks) that the word *robot* comes from the Czech *robota*, meaning servitude or forced labor or drudgery. In other words, robots have generally been understood to be machines that perform some kind of human labor, typically associated with being repetitive or difficult labor.

That's part of the equation, understanding that robots are like the washing machines and dishwashers and combustion engines of the past century. They're machines intended to make our lives more efficient. Whether or not they actually achieve this is another question altogether, but here's the point: when it comes to the humans creating these robots, it's about innovation. The only way we'll "die" is if we fail to innovate.

Fortune has recently created a new category called the Fortune Future 50. Their methodology weighs 17 predictive factors in the areas of strategy, structure, technology, and people. Among these factors are a company's board composition, its environmental, social, and governance standards, and the share of managers and employees who are female.

Fortune's algorithms utilize AI to assess a company's expected future growth and its capacity to deliver that growth. That's how the future is being anticipated, not from those who want to romanticize the past, but from the realists who are ready to be nimble and creative and industrious.

Robots are not replacing us, not anytime soon. For one, they can't innovate. They can't interpret the data we've programmed them to create. And for the time being, they still can't drive. That's a topic for another decade or two down the road anyway.

Before we get there, we face a nearer-term future.

I think of Steven Johnson's concept of "the adjacent possible." The adjacent possible is a kind of potential future, hovering on the edges of the present state of things, a map of all the ways in which the present can reinvent itself. Johnson has written that the adjacent possible "captures both the limits and the creative potential of change and innovation." It's the ever-present set of opportunities at the boundaries of our reach. "The strange and beautiful truth about the adjacent possible is that its boundaries grow as you explore them. Each new combination opens up the possibility of other new combinations."

According to Johnson's idea, what I see is the complementary role robots will play in our lives. We're going to work alongside them more and more. We already are. What I see happening is that rather than getting replaced, we're figuring out ways to increase efficiency in our lives, and we are innovating areas that humans either can't or don't want to do — at least not as well or as long as robots.

Some people are actually saying there's no way consumers can continue to expect to get their free shipping in 1-2 business days. They say that e-retailers have created unrealistic expectations.

The costs are simply going to skyrocket, they say, whether it's from our inability to contain rising freight and capacity costs, or because of rising fuel costs, or because our infrastructure just won't support the ever-increasing density of so much e-commerce transport.

Here's why I think they're wrong.

Just like the internet disrupted the 2000s and social media the 2010s, we're looking at automation in the coming decade. But are we really staring down the barrel of the singularity, when the machines grow advanced enough to make humankind obsolete? I think not.

The more realistic scenario is that humans and robots are poised to live in harmony. This is the idea of multiplicity. You're more likely to work *alongside* a robot than be replaced by one. If your car has driver-assist cruise control, you're already letting the robot handle the dull highway work while you take over for the complexity of city driving.

Should machines help us drive? Absolutely.

While autonomous startups have faced a litany of setbacks over the past year, the obstacles they face are immense. At the same time, they're continuing to innovate and invest enormous sums of money into solving this logistics problem. Whether or not it's the startups or a giant OEM incumbent that decides they're ready to invest even more billions into the solution, it's happening.

There are of course industrial robots created for heavy-duty automation for dangerous jobs, which have been common in auto production since the 1960s. There is also a lot of recent action in the robotics venture capital space, especially along the lines of cobots.

These are sensor-laden robots that can work alongside humans without cages. Also, we're seeing their emergence in warehouse spaces. They're autonomous guided vehicles (AGVs) for material transport within the warehouse or shop floor.

While the startup scene is also investing heavily in developing sensory and control solutions for robots, the main idea is for them to work alongside humans — not replace them.

We really aren't anywhere close to some kind of *Bladerunner* dystopian future, where the bad Replicants are needing to be retired, but at least the good Replicants are obedient.

One of the fundamental challenges of developing any kind of robotics program hinges on Moravec's Paradox: "It is comparatively easy to make computers exhibit adult-level performance on intelligence tests or playing checkers, and difficult or impossible to give them the skills of a one-year-old when it comes to perception and mobility."

Our focus at FreightWaves is on developing the AI. Where is the intelligence source if not from the data we input? That's what we're doing behind the scenes: we're aggregating data. Data is the new oil for the very reason of automation.

At FreightWaves we're creating a technology market platform for freight transportation, a futures exchange.

Freight transportation is a huge component in the costs of goods that everyone consumes, and like energy and raw commodities, freight costs are extremely volatile. Between January 2016 and January 2018, US trucking rates were up more than 25 percent.

In North America, trucking is the largest mode of freight transportation, representing more than 73 percent of all freight transportation services.

The reason that trucking rates are so volatile is as simple as Economics 101 — supply and demand. On any given day, there are a finite number of trucks in the market, and there are a finite number of loads in the market. Truckers adjust their rates based on current market conditions and competition for capacity.

In markets that are volatile, participants have significant exposure to their cash flows as prices move up or down. If the market is big enough, futures contracts are established to help participants de-risk their exposure to volatility by creating hedging instruments that they can use to have more predictable cash-flow streams.

In addition to providing a risk management option through hedging, trucking freight futures contracts will help bring transparency to the industry by giving market participants a sense for where rates are headed in the near-and-long-term.

Once the trucking freight futures contracts start to trade, a forward curve will be visible, which will help participants understand the market's expectation of forward/contract rates. This will help them plan their operations, make better capital expenditure decisions (should they buy trucks now or not), and price freight more accurately to the market.

We're building the machines, giving them the intelligence that will pull from hundreds of sources when it's all said and done.

That is a form of knowing that far exceeds the power and analytical ability of any one human.

It has been—and will continue to be—a collaborative effort. In other words, we're building our own A.I. People can access this intelligence for their myriad perspectives—whether they're taking a long or short term outlook, whether they're a carrier, broker, or shipper, etc.

They and others will also use their own analytical tools to consider and interpret the data. Humans working alongside robots, and vice-versa. That's where I'm hedging my bets, on a future that's emerging right out my windshield. ~

Craig Fuller is the CEO of Freightwaves, where he is creating a futures and options market for transportation spot capacity and working toward providing industry participants with a transformative new way to protect against price volatility on major US trucking routes.

Construction Discovers the Meaning of Robotics Innovation

Kaleb Steinhauer

- CEO of Genesis Dimensions -

The construction industry is evolving. Companies have started to realize that they must embrace technology to be a part of the future landscape in one of the world's largest industries.

About three years ago, when I started my company, Genesis Dimensions, a lot of companies in construction didn't even know what innovation meant. Now, almost every construction company has taken the first step and installed some sort of an innovation team to help them plan for the future.

The only problem with this is that some of them can't even spell the word *innovation*.

Of course, these innovation teams have the potential to enact change, and they are a great first step. But there are some limitations to what they can accomplish.

It's sort of like planting a pecan tree; while it may grow and it has branches, it's going to take a few years until you get something in return.

Most of these teams are led by a head engineer who is very risk-averse and is more focused on technology ideation than business-oriented initiatives. So, these innovation teams often do a lot of thinking and planning, but they don't actually know what to do yet.

This means that while there is a lot happening behind the scenes, it will be some time before the average person will recognize the difference in their lives. However, accepting that there is a need for innovation is still a critical first step for the construction industry.

This thought movement will help to create a greater innovation ecosystem, where people and entrepreneurs start to see a path toward funding an industry that is willing to start trying. Venture capital firms are beginning to focus on hardware solutions because they see the long-term potential.

The Department of Defense just launched a committee in partnership with the American Concrete Institute to work on 3D printing. And this is only the beginning of the integration of technology into the industry.

In the next two to five years, I think there will be a significant acceleration of technology adoption that could help change the perception and trajectory of the entire industry.

Nationwide adoption will take time, but people are working hard to make this happen.

One hurdle construction startups must overcome is that there is a bit of wait time when it comes to profiting off hardware. There is a lot of research and development that startups must do before they have a commercially viable product. The return on investment in the industry has the potential to be huge, but it may take longer than a software-focused startup to deliver.

Some other challenges that construction innovators will face in the future are regulatory, public perception, or safety.

First, government regulation is a massive issue when it comes to construction. Every building, road, and system must be constructed to certain specifications to be approved.

Making sure that new technologies can work with these standards will be critical for success. Second, changing the public's perception of the construction industry will also be essential to achieving commercial viability. Most people currently don't view the construction industry as particularly efficient or effective.

I think, with greater technology adoption and implementation, the industry has a chance to change people's minds with new and innovative methods. The final and biggest hurdle the industry must clear with new technology will be safety.

This ties into regulation somewhat, but people need to understand that construction is a dangerous industry where human lives are at stake. While implementing a new technology, you not only have to prove that it is useful and improves the process, but you also need to make sure that it is safe. This is for both worker safety and for the safety of the people inhabiting or using whatever you are building.

In the next year, I think there will be some surprises for robots and automation in the construction industry. We may also see new subindustries created within the construction industry that will be needed because of the emergence of new technologies. We will also see government bodies take a more active role in trying to understand how these new technologies will change the industry.

In 2019, while there will be heavy investments in R&D and market adoption, I don't foresee any massive changes that will result in something that could cut costs by 50 percent. But I think in a few years that might be the case. But right now, the industry has just been conceived, and the birth hasn't happened yet. ~

Kaleb Steinhauer is the CEO of Genesis Dimensions. Kaleb's solution combines large-scale 3D printing and robotics to revolutionize the construction industry through lower labor costs and limited environmental impact.

From Smart Pilots to Smart Cities

Fred van Beuningen

- Managing Partner of Chrysalix Venture Capital -

Technologists have promoted "smart cities" as utopias where webs of sensors, artificial intelligence, and robots will cure the woes of urban life. The only problem: no such cities exist yet.

What is the state of the smart city, and how will it evolve in 2019? These questions matter because 600 cities already account for 60 percent of global GDP according to McKinsey.[1] Moreover, the UN predicts that 66 percent of the world population will be urban by 2050, up from 54 percent today.[2] Smart cities will be habitats for most of humanity. Investments we make now will influence global quality of life for decades.

In 2016, Statista counted 252 smart city projects worldwide.[3] A few of these examples can help us (a) understand imminent changes in city living and (b) model how automation and robotics will affect urban development.

Smart city pilots in Dallas and Seoul demonstrate that improved energy efficiency and waste management are within reach of any city. Amsterdam illustrates how cities might connect smart energy and waste management projects to create looped use cycles. Finally, pilots in Helsinki and Toronto suggest how one loop of interconnected smart systems could scale into a "smart city" deserving of that moniker.

In 2019, I expect public and private entities to integrate isolated, departmental services into interconnected, looped processes. The cities that do so will turn smart pilots into smart cities.

Isolated Intelligence: Lighting in Dallas

In March 2017, the nonprofit Dallas Innovation Alliance (DIA) established the "Living Lab," a four-block corridor where 30 organizations could prototype a smart city.[4] DIA launched nine projects focused on lighting, water use, connectivity, parking, and other city needs. The project was collaborative in mindset but not in implementation.

The standout project was intelligent LED lighting, which reduced energy use by 35 percent during a 12-month pilot. Applied citywide to 85,000 lights, DIA calculates that intelligent lighting would save $90 million over the life of the LED bulbs. How? DIA says that the lights were managed by "intelligent nodes, which allow for remote lighting management, outage alerts, and operational efficiencies."[5]

Lighting, argues Navigant Research's Eric Woods, is the "backbone" of smart cities because it can host myriad systems.[6]

For instance, the Dutch company Sustainder adds cameras as well as sensors for movement, weather, carbon dioxide, and sound to its intelligent lighting system.[7] To a Sustainder lamppost, a city could add a Danish-designed Leapcraft sensor, which monitors pollutants and particles.[8] A series of such lampposts could ingest data for intelligent parking, public safety, traffic management, Wi-Fi provision, and more.

If DIA connected intelligent lighting to the eight other projects, they didn't mention it. Living Lab's lighting system, though efficient, is isolated. Dallas has smart pilots, but not a smart city. That's wasteful, as the next sections will argue.

Seoul's Garbage Problem

Seoul, the world's second largest metropolitan area by population, had a severe litter problem. The city's trash bins, insufficient in size and quantity, overflowed regularly, says a case study by the Smart Cities Council.[9] Recycling was rarely practiced. Even if city goers tried to use the bins, the Public Cleanliness Department could not collect the trash frequently enough.

What does waste in Seoul have to do with lighting in Dallas?

In 2014, Seoul installed 85 smart trash bins made by Ecube Labs.[10] The Public Cleanliness Department monitored the fill levels and optimized collection routes to prevent overflow.

These 85 bins eliminated overflows with a 66 percent reduction in collection frequency, 83 percent reduction in collection costs, and 46 percent increase in recycling diversion rates. Smart technology succeeded, just as it did in Dallas.

Yet once again, the bins are an isolated system independent from other city services. Although waste no longer overflows, it still goes to landfills. After all of Seoul's effort to collect waste, is there not a better use?

Making Light of Waste

Dallas and Seoul each have a smart project that works. In total, 252 cities do too, but we don't call them "smart cities," per se. The smart city is not the result of one good project or even the culmination of spreading nine good projects from four blocks to 50. Rather, a "smart city" loops systems and resources into one another, forming one network. At that, Amsterdam excels.

The Amsterdam Smart City Project launched in 2009 with an unusual starting point. The city invested in power, connectivity, and data on behalf of the founding partners and then welcomed "bottom-up innovation," say researchers at the University of Palermo.[11] The result was more than 20 pilot projects and collaborations with 72 partners in the first three years. Whereas most smart pilots have a panoply of short-term goals, Amsterdam focused on one: reduce carbon emissions 40 percent by 2025 (compared to 1900 levels) and 75 percent by 2040.

Amsterdam's landmark project is to turn waste and sewage into sustainable power. In 2017, the Palermo team reported that Amsterdam's waste-fed power plant produced 560 GWh of electricity per year, more than enough to power the tram and underground, town hall, and public lighting. The waste also generates 548,000 GJ of heat for homes and businesses.

Meanwhile, Amsterdam's purification system separates drinking water from sewage sludge, which produces 11,000,000 m^3 of biogas when fed into an incinerator. In turn, the biogas provides electricity and heat for water purification.

The circularity of this system is a model for other smart cities, including Dallas and Seoul. A smart waste collection system like Seoul's could fuel a waste incinerator like Amsterdam's to power a lighting system like Dallas's. In 2019, such a loop is realistic — and merely a starting point.

Circular Automation and Robotics

The Dallas, Seoul, and Amsterdam examples propose that a smart city, using intelligent sensors and data systems, could collect waste, convert it to sustainable energy, and thereby power efficient lighting systems, which form a backbone for the smart city. Those steps, until recently, demanded manual processes and people to perform them. Now, thanks to advances in mobility and robotics, we can and should automate the circle in 2019.

The question is not if we can do this, but how we should do it. Compare Seoul's waste solution to one Helsinki implemented in its Kalasatama smart city district. There, residents separate their waste into mixed, bio, paper, and cardboard bins and then the system "whooshes" the trash at 70 km per hour through subterranean tunnels.[12] It reaches a central collection station, where trucks collect the waste for further handling. That results in 80 to 90 percent less garbage truck traffic.

Now, compare Seoul and Helsinki to Sidewalk Labs, a pilot in Toronto that plans to create subterranean tunnels in which robots transport mail and garbage via underground tunnels.[13] Which of the three options is the most scalable for a city?

Seoul could use autonomous garbage trucks to gain efficiencies immediately without digging tunnels. Helsinki's tunnels might be slower to implement, but they automate collection without robots, saving resources for robotic processes in, say, a waste power plant. Sidewalk Labs makes dual use of the tunnel network (and eliminates noisy delivery drones overhead), but it demands upfront investment in robots.

Perhaps the choice of waste delivery will depend on the city's resources, talent base, and geography. The important thing is to automate trash delivery to a power plant, which, in turn, electrifies the delivery system and the intelligent lighting above. That lighting system might also monitor weather and foot traffic if it's made by Sustainder and monitor pollutants and particulate matter if we attach a Leapcraft sensor. If we build upon basic loops, we might see the first "smart city" in 2019.

From Smart Pilots to Smart Cities

The world's 252 smart pilots must integrate siloed projects into circular systems. No metropolitan area deserves the title "smart city" until it is interconnected that way. Semi-intelligent cities can and should make that leap in 2019.

The energy-waste cycle, the focus of this chapter, shows how existing, proven technologies would form a circular system. It also demonstrates the interdisciplinary nature of smart city automation and robotics. Innovators in disparate fields — like energy generation, waste management, and autonomous vehicles — must form partnerships. No single private or public entity can build the circular smart city alone.

For 2019, I have a pitch for smart city innovators. From proven smart city pilots, let's create one loop. From one loop, let's create entire smart cities. ~

ENDNOTES

1. https://www.mckinsey.com/featured-insights/urbanization/urban-world-mapping-the-economic-power-of-cities
2. https://esa.un.org/unpd/wup/publications/files/wup2014-highlights.pdf
3. https://www.statista.com/statistics/801167/project-of-global-smart-cities-by-region/
4. http://www.dallasinnovationalliance.com/
5. https://www.bizjournals.com/dallas/prnewswire/press_releases/Texas/2018/11/08/UN66280
6. https://www.forbes.com/sites/pikeresearch/2018/04/06/smart-cities/#2347d43913c8
7. http://sustainder.com/en/

8. http://www.leapcraft.dk/
9. https://smartcitiescouncil.com/resources/case-study-city-seoul
10. https://www.ecubelabs.com/
11. https://www.researchgate.net/publication/310485601_Smart_Cities_Case_Studies
12. http://smartnclean.helsinkibusinesshub.fi/projects/automated-waste-collection-system
13. https://www.sidewalklabs.com/

Fred van Beuningen is Managing Partner at Chrysalix Venture Capital. Fred's industry expertise spans C-level, Executive Board and CEO positions in oil & gas, chemicals, logistics, innovation and sustainability, and scaling up ventures; he is based in the Netherlands.

Blockchain and Automation in the Financial Services Sector

Robert Handfield

- Professor at NC State University and Author -

Much ado is made about how there is a myriad of automations emerging, including blockchain, process automation, AI, and digitization, and an entirely new wave of technologies coming at us. This next wave of technologies has the potential to transform trading relationships and networks. But to do so, will also mean that we must begin to move toward a high level of standardization — one that allows technologies and systems to communicate with one another.

Standardization of data has driven tremendous improvements in the ability of different systems to communicate with one another, to a point where standardization will lead to the disaggregation of activities away from enterprise to a diverse group of companies that will manage such outsourced services.

In financial services, an extreme amount of due diligence has been given to internal controls, but as these organizations have outsourced more and more, they do not have the same level of insight into quality and performance of diverse suppliers and customers as it is beyond the capability of existing technology.

As a result, organizations are now having to throw a high level of human capital at the consequences of outsourcing to manage these resources. The following examples provide insights into how these changes will occur.

Dynamic Discounting

New technologies will have a big impact in the future landscape of financial flows in the supply chain. At a large global bank, for instance, procurement is employing a dynamic discounting tool that provides opportunities for cost saving as well as early payment discounts for suppliers that require access to working capital. For example, typical payment terms may be 10 days with a 2% discount, or zero discount in 30 days.

But what if the supplier needs cash immediately?

What would they be willing to pay to get the payment within two days?

Some suppliers may even want to factor their payables to access letters of credit and are willing to pay higher interest rates on such commercial paper.

This bank is piloting a tool that enables supply chain managers to set an internal hurdle rate and put current payables out on an auction. The auction will allow suppliers to put their payables out to bid, and automated bidders will provide different interest rates that they are willing to cash in the payable. The supplier can choose among these options and take the lowest rate to factor their payable. This can yield significant cost savings for the bank and help suppliers with their cash flow positions.

Blockchain

Much has been said about blockchain, or distributed ledgers. This same bank has a pilot with a large outsourced systems provider and uses smart contracts to enforce service level agreement (SLA) management of their server contracts within their outsourced data center. The systems provider manages 400 prepositioned servers for the bank, which is in their data center.

Each sub-server has an Internet of Things (IoT) sensor, and is pre-positioned for the bank, and is tied to a condition in the contract for those storage services. When the server is provisioned, the IoT trigger automatically triggers a purchase order and invoice for that server, and once provisioned, there is a Service Level Agreement (SLA) that goes into effect, stating that the server must operate at an uptime of 98.6%, for example.

A payment goes to the service provider upon the server being provisioned, which is made in the bank's own cryptocurrency.

However, what happens if the server fails due to a faulty chipset?

If the server fails to hit the SLA, two things happen.

First, the server sends a signal into the blockchain, and it generates a service credit. Second, the blockchain immediately creates a settlement for the credit, using the cryptocurrency created by the bank (let's call it "bankcoin").

Bankcoin is an internally generated cryptocurrency, with a defined value of one dollar.

The potential future scenarios for applications of blockchain are enormous for the financial services sector. One of the biggest expenses for banks is professional services, and one of the biggest challenges associated with professional services is how to monitor services delivered in a Master Services Agreement (MSA). From a contract management perspective, the challenge is how to ensure that the verification of services performed occurs against the contracted requirements and statement of work.

For instance, if using an "agile" software development perspective, banks are thinking about how they could build a two-week deliverable checkpoint into the process, which would allow the customer to send a signal back into the blockchain showing that the code had been completed, tested, and verified.

This would in turn be used to drive the payment cycle.

This is one of many potential future applications that would transform procurement and potentially eliminate the need for procurement software that requires processing of POs, invoices, requisitions, etc.

The coming year will be an exciting time indeed for new technologies in the financial sector. Many more opportunities abound!~

Robert Handfield is the Executive Director of the North Carolina State University's Supply Chain Resource Cooperative. The SCRC is a university-industry partnership dedicated to advancing the supply chain industry and the professionalism of its practitioners.

Robots in the Supply Chain

FI THE FUTURIST INSTITUTE

Key Supply Chain Technologies to Watch

Daniel Stanton

- President of SecureMarking -

In 2019 the headlines are going to be filled with stories about how automation and robotics are transforming our lives and our supply chains. These stories are likely to highlight three important themes: technology innovations, business impacts, and human impacts.

We'll see many examples of amazing new technologies that enable automated machines to perform tasks with superhuman strength or speed. We will also see the many ways in which automation is increasing productivity and improving the bottom line for companies. Finally, these stories will show that while automation can improve efficiency and benefit customers, it will also eliminate some jobs and make many employers seem cold and inhuman.

For now, people remain critical to supply chains.

We are unlikely to see a "lights-out" warehouse or a fully automated supply chain. To understand why, you should keep an eye on the evolution of four technologies that are key to enabling automation in the supply chain: blockchain, artificial intelligence, autonomous vehicles, and additive manufacturing.

Let's look at the current state of each of these technologies, using the following phases to describe the maturity of these technologies:

Technology Adoption
Conceptual
Prototypes
Pilot Projects
Niche Adoption
Strategic Adoption
Widespread Adoption

Blockchain	Pilot Projects
Artificial Intelligence	Strategic Adoption
Autonomous Vehicles	Mixed
Additive Manufacturing	Niche Adoption

Graphics: Provided by permission from Daniel Stanton.

Blockchain (Pilot Projects)

Right now we are seeing the preliminary use of blockchain in a handful of supply chains where it is unlikely to have a major impact. That's because everyone wants to see how it works, but no one trusts the technology enough to bet their supply chain on it. Throughout 2019, we'll hear stories about tracking food and diamonds, and sharing information about international shipments. But at this point, the blockchain is likely to be an add-on, not a mission-critical part of these supply chains. While the security and reliability of blockchains are assumed to be high, it is also a fair bet that we'll see some high-profile hiccups.

For example, we may begin to see gaps between the promise of perfectly transparent data stored in a blockchain and the murky realities involved in moving physical products around the world. And we may realize that businesses prefer not to have permanent records of every transaction because this can actually increase their risk for future litigation. But the biggest concern is whether we'll see viruses or vulnerabilities emerge that allow hackers to hijack a blockchain for nefarious purposes.

As a practical matter, today's business conversations often get bogged down trying to explain how blockchains work. The key point is that blockchain is a different way to store and share data than what we've done before, and it has advantages in terms of privacy, data integrity, and resilience. As some of the market leaders, like IBM, are beginning to prove those advantages out through high-visibility projects, we'll start to see blockchains replacing traditional databases in critical legacy applications and enabling new business models and processes.

Artificial Intelligence (Strategic Adoption)

More and more devices are taking advantage of machine learning, ubiquitous connectivity, and cloud-based storage and processing to make decisions. We are gradually coming to accept these systems in our daily lives, but the technologies are imperfect in their perception and in their decisions. And of course, different AI systems have different capabilities. Therefore, these systems are often relegated to decision support roles, with human fail-safe steps built into their processes.

For example, if you tell your iPhone, "Siri, call Robert," it is likely to respond with a check question, such as, "Do you want me to call Bob?"

In 2019, most supply chain managers won't be willing to pass control of important supply chain processes over to a machine. But companies like Alloy have shown that there can be benefits in using AI to oversee the vast amounts of data in supply chain information systems. In particular, these tools could transform the way we manage supply chain planning, execution, and visibility.

Autonomous Vehicles (Mixed)

Supply chain robotics appear to be following the path of disruptive innovation described by Clayton Christensen in *The Innovator's Dilemma*. They are taking on low-value, mundane activities like moving and lifting products — tasks that are unattractive to human workers.

In this sense, these robots are helpers or collaborators, and many robotics firms are now using the term *cobots* to highlight the positive aspects of this human-robot relationship.

The rising costs and lower availability of workers will continue to be strong drivers toward automation in manufacturing and distribution operations. And the rapid evolution of hardware and software will continue to make these tools more powerful and less expensive.

A short-term risk to automation is the potential for rising interest rates and recession, both of which will make capital investments less attractive. A significant economic downturn could lead to projects being canceled or postponed, and that could put many automation companies into financial jeopardy, leading to a wave of consolidation in the industry.

There is a lot of variation in the maturity of autonomous vehicles. (Note: my emphasis here is on autonomous vehicles that are used in supply chains, as opposed to those used for passenger travel or military applications.)

This is why it is helpful to categorize autonomous vehicles by mode and intended travel distance, as in the chart below. Some combinations of mode and travel distance are impractical (delivery by water within the four walls of a facility, for example). But within the viable categories, the development of autonomous technologies is uneven.

To illustrate, I've included the names of some of the companies that are leading the development of autonomous vehicles in each category.

Adoption of Unmanned Vehicles in Commercial Supply Chains				
	Ultrashort Range (Four Walls / Campus)	Short Range (Last Mile)	Mid Range (Trans / Interstate)	Long Range (Trans / Intercontinental)
Aerial	Strategic Adoption (Dronescan) (PINC) (Walmart)	Niche Adoption (Zipline) (Matternet)	Conceptual (Lockheed Martin) (Flyox)	Conceptual (Lockheed Martin)
Above Ground	Strategic Adoption (Amazon) (XPO) (DHL)	Pilot Projects (Starship) (BoxBot) (Marble)	Pilot Projects (Tesla) (Embark) (Kodiak)	Pilot Projects (Rio Tinto)
Underground		Pilot Projects (Volvo)	Prototypes (Hyperloop)	
Above Water			Prototypes (DNV GL) (Hrönn)	Prototypes (Rolls Royce)

Graphics: Provided by permission from Daniel Stanton.

Additive Manufacturing (Niche Adoption)

Additive manufacturing, or 3D printing, is being used in niche applications that require expensive machines to create parts out of a homogenous material in central locations. These criteria align with some industrial components, such as aircraft engine components, but they exclude most consumer products.

3D printing has become a ubiquitous solution for applications including rapid prototyping, parts with complex geometries, and small-scale production. But the broader adoption of distributed manufacturing — for example, a neighborhood 3D printing shop — will face a number of practical challenges, ranging from the need to use multiple material types to implementing quality control and intellectual property protection.

Summary

Blockchain, artificial intelligence, autonomous vehicles, and additive manufacturing each play a critical role in the development of automated supply chains. Each of these enabling technologies needs to be developed further, and they are all advancing quickly. But to truly understand the role that they will play, we need to monitor the technological capabilities along with the business and social impacts. Someday we will have lights-out warehouses and fully automated supply chains. But in 2019 humans will still be essential for efficiently delivering goods and services around the world. ~

Daniel Stanton is the author of *Supply Chain Management For Dummies* and several popular supply chain courses on LinkedIn Learning. He is a co-founder of SecureMarking, which has developed a robust solution to address the growing threat of counterfeit, black market, and gray market products in global supply chains.

Time for Warehouses to Ride the 4.0 Wave

Samuel Alexandersson

- Product Marketing Manager of Kollmorgen -

The booming e-commerce sector is pushing suppliers to rethink the movement of goods throughout the supply chain and helping fuel the fourth industrial revolution. Automation is ensuring that the growing demand for faster deliveries is met, and this is giving rise to an increasing number of smart warehouses around the world.

Now you can kick-start your own revolution with automated guided vehicles (AGVs) — and ride the fourth wave of industrialization that's sweeping the world. 2019 will be an inflection point, separating the digital winners from the laggards.

So, if you haven't already begun experimenting and checking your current digital competence, now is the time — otherwise you risk falling behind.

Full Steam Ahead on All Fronts

On Singles' Day in 2018, an annual 24-hour buying frenzy that started as a celebration for singles in China, e-commerce giant Alibaba registered more than 1 billion transactions, resulting in an all-time-high daily sales record of $30.8 billion, up 27% since last year. As astounding as these numbers are, what is taking place inside warehouses is equally astonishing. It is here that the fourth industrial revolution is now in full swing, leading to more and more smart warehouses. Automation plays a key role in this transition, ensuring fast, smooth, and flexible deliveries — critical to the modern buying experience.

These numbers may seem exceptional, but considering that China has more than 800 million internet users, of which 98% are mobile, ordering things online has become a natural part of everyday life. In today's mobile-first world, people expect to have goods and services delivered to their doorstep after a few clicks on their phone.

However, as many as 80 percent of warehouses are still manually operated. Since consumers can now purchase anything, anywhere, and at any time online, optimizing your logistics is key to keeping customers happy and your business growing. Implementing an automated solution promises not only higher efficiency and safety inside your warehouse, but the opportunity to give staff more inspiring jobs, leaving boring and hazardous tasks to machines.

Paving the Path to a Smart Future

The Boston Consulting Group listed self-driving logistics vehicles, or AGVs, as one of nine foundational technologies, driven by the need for automation, that are decreasing costs and increasing productivity and growth.

Today, AGVs come in various sizes and degrees of customization. Vehicle types include unit loaders, platform carriers, assembly platforms, and lift trucks. For decades, automated vehicles have minimized the need for manual labor in dangerous, dirty, and demanding environments as well as freed staff from tedious and repetitive tasks. However, this is no longer enough. To build and sustain a lead in the race toward the fourth wave of industrialization, you must deepen your digital knowledge.

In the past, industry was often dependent on locked processes for machines and conveyors. If you have an AGV or mobile robot, you can tell it where it should be, how it should act, and what it should do.

AGVs are often used as parts of a system to perform different processes. The vehicles provide data, and with this information, you can optimize the system and — using AI — make it self-learning over time. The result is future-proof intralogistics.

This kind of flexible automated solution is a way to handle a changing world. Consumer behavior is evolving, raising the demand for flexibility in logistics. AGV solutions fulfill many of these demands and have a natural place in this new world. They are flexible in that they can adapt to any infrastructure. Sensors enable them to follow different paths. These paths are controlled by computer software for the vehicle fleet that is integrated into the plant's warehouse management and business enterprise system.

Although AGVs may not be the answer to all your Industry 4.0 challenges, they respond to a surprising number of them. With this future-proof system, managers can look forward to years of profits, a safer workplace for staff and reduced waste. Implementation is relatively easy, meaning you can be up and running in just a few weeks. The payback period is also short, often as low as 1-2 years.

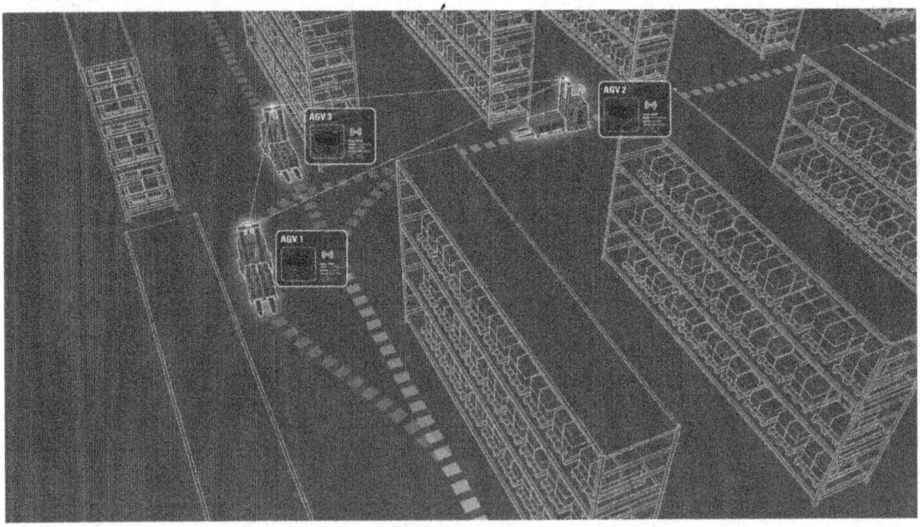

Getting Everyone on Board

You can rest assured that AGVs will improve warehouse efficiency as well as the well-being and safety of employees. Fast-reacting sensors prevent collisions between people and other vehicles, allowing people to work alongside these automated vehicles. AGVs are also proven to reduce plant noise and exhaust fumes, resulting in a more pleasant and cleaner working environment.

But before you implement an automated systems solution, you need to ensure that staff understand these benefits. This requires careful change management. People may fight the change in fear of losing their jobs. But contrary to what many believe, automation creates new and more engaging jobs. A study by the management consultancy firm Boston Consulting Group shows for example that automation will result in 350,000 new jobs in German manufacturing industries alone by 2025.

How to Respond When the Future Calls

With the e-commerce sector growing by 10 percent yearly, it's only a matter of time until most warehouses are automated. To stay competitive, forward-thinking companies are already undergoing major change inside their facilities and automating their operations. AGVs have and will continue to withstand the test of time, offering a proven and flexible solution. Now, you too can make your warehouse smart enough for the future and ride the wave of opportunities it promises.

The constant challenge of meeting the changing needs of the marketplace is what drives me forward. Business is booming and my role of finding opportunities and then addressing them with insights is highly rewarding. My main job is to analyze and join the dots to successfully segment and position a strong portfolio of products for the future market. A rewarding source of motivation in this global environment is all the great people you meet. Not only colleagues, but stakeholders from all over the world. ~

Samuel Alexandersson is the Product Marketing Manager at Kollmorgen. He holds degrees in Industrial Management and Engineering from The Faculty of Engineering at Lund University.

Hyperlocal Supply Chains: Physical, Financial, and Available

David Schwebel

- VP of Business Development and Market Intelligence -

Hyperlocal online retail businesses are popping up in dense urban areas all over the world. I've been calling it the "mom and pop" reaction from big online retail firms. Let's break down what a hyperlocal e-commerce business is and what we can expect to see in the next 12 months.

What do you mean by hyperlocal e-commerce?

Businesses deploying hyperlocal e-commerce fulfillment channels want to provide same-day goods and services for all customers and consumers usually within a 2-hour drive radius of their closest inventory point (e.g. store, distribution center, or 3PL fulfillment hub). Think same day purchase & delivery of food, medicine, consumer goods, retail products, and services that are valued on convenience and not tied to just your home delivery.

In general terms, *hyperlocal* refers to a very specific area, area in proximity of your home or your business or your current location. The term *hyperlocal commerce* arrives with the addition of payments and marketing options available in your consumers' vicinity.

So, even though the local coffee shop five blocks away from you doesn't deliver, "Joe Quick" on Postmates, UberEats, or a more localized service will bring coffee right to your office or couch for a nominal fee.

While that cup of joe is inbound, why not have "Joe Quick" pick up windshield wipers for you from AutoZone or binder clips from Office Max for a few extra bucks?

The hyperlocal business operates within a local proximity to high population and retail density. Did you know that more than 50% of the US population can be reached within 2 hours' drive time? That's why global retailers are investing in this new fulfillment channel.

What is the attraction to hyperlocal?

Speed. Hyperlocal businesses can deliver products and services faster than larger online retailers. Hyperlocal food delivery time is already less than an hour on average. Products you would normally have to order by a certain time to receive next-day service are now available to you within a matter of hours. Additionally, most of these deliveries are based on your exact vicinity and not solely reliant on a street address.

Trust. The consumer is put in a position where they can see and understand all the steps of the supply chain from purchase to sourcing to delivery. Consumers know where the product is made and most likely who will be delivering it to them. The hyperlocal businesses have a huge incentive to capture and retain customers by building that trust and enhancing the overall customer experience.

In layman's terms, the meat tastes better when you know the butcher and the butcher knows your cut. This is why speed and trust will enrich consumer word of mouth, becoming a key component for hyperlocal business success in their communities.

Can I get an example? The 'hyperlocal' grocery

"Little Joe" survives on pizzas, burgers, and video games. He hasn't visited a department store in months; everything he wants can be ordered online. Ironically, he just found out that his parents will pay him a visit tomorrow; he has no other option but to visit a nearby grocery store to buy at least basic kitchen necessities.

Yet he just saw an ad for Kroger's Click & Pick and/or Target's Shipt, so he downloaded one of those apps, registered on it, and started ordering groceries online. He picks what items of grocery he needs, makes the payment in the app itself, and waits for the order to arrive.

Fortnite doesn't play itself, you know.

In the meantime, the app owner passes the grocery list to the delivery partner, begins processing that order in real time via some really advanced automated fulfillment technology, signals the order picked complete, assigns the order to the next available last mile delivery route (owned by the distributor or third party), and delivers it to Little Joe. All in two hours or less.

While this sounds simple to implement, it is easier said than done. It has its own set of challenges.

An extensive 10-year research study from Chalmers University of Technology in Sweden provided a comprehensive look into the material handling systems that would be used in hyperlocal e-commerce warehouses. They identified that the traditional approaches to warehouse automation rely on mechanized conveyor and sortation systems that aren't easy to install, maintain, or scale. Especially in dense urban environments, solution sets that focused on the entire value intersection required the flexibility of robotic-enabled fulfillment. Fitting into any sized building, including ceiling heights of 9 feet or less!

Robotic automatic storage and retrieval goods-to-person systems like Autostore scored +21 — fast, scalable, and dense fulfillment in both ambient and refrigerated environments, quickly becoming the "vending machine" for hyperlocal fulfillment.

Rack-carrying AGVs like Carrypick scored +25 — dynamic, scalable, redeployable, goods-to-picker technology that acts as the "any-sized" product engine of hyperlocal fulfillment.

When most people think of robots, their minds tend to go to either the fixed robotic arms that consistently perform repetitive tasks in manufacturing or humanlike robots from Boston Dynamics doing a CrossFit routine.

Neither of those types of robots has a place in the dynamic hyperlocal warehouses — at least not today. One lacks the mobility and finesse fulfillment requires, the other the functionality.

What we're talking about are highly functional, compact robotic systems created to perform specific tasks in the warehouse, such as consistently and reliably delivering products to workstations where humans can efficiently package them for delivery.

By having robots find and move specific products in this goods-to-person model, it's no longer necessary to have humans walking up and down aisles all day pulling orders. This not only makes better use of human resources, it results in dramatically reduced pick times.

What's next?

The rules of hyperlocal e-commerce fulfillment are still being written. As we see changes to urban landscape and infrastructure, it will become easier for larger e-commerce retailers to integrate into our megacities with the advancement of hyperlocal fulfillment centers empowered with robotic automation that is available today.

Most companies that enable this fulfilment method will now need to increase their outreach, trial, and adoption efforts to educate the consumer public that an intraday delivery is more desirable then waiting days for that delivery, and that it will provide a brand-remarkable experience versus their competition.

Known in poker as "table stakes," the advancement of consumer awareness in retailer supply chains will rapidly require these hyperlocal fulfillment centers to become a requirement of global value chains, further rewarding all in our ever-evolving ecosystem of speed and trust. ~

David Schwebel is the VP of Business Development and Market Intelligence at Swisslog Warehouse & Distribution Solutions, where he helps forward-thinking companies optimize the performance of their supply chains with future-ready automation systems, robotic solution sets, and innovative software.

Skill Gaps and the Multi-Axis Bandit

Jayesh Mehta

- Marketing Manager of Transbotics -

As we embark into 2019, a few major changes have happened in the robotics and material handling industries. Some are facts-of-life-type changes, others spark intense arguments and emotional decisions. I'll start by talking about the controversial, because I like causing a little chaos; it adds spice to life...at least for me it does.

Whether you agree with the US leadership or not, there are some short-term results that few could argue. The main one is the increase in production by many midsize to large companies. This has caused the amount of job openings to increase, which in turn should theoretically benefit the economy. This really becomes a mixed bag when you delve into what jobs are created, the pay scale, and the amount of jobs that do not get filled. Most of the jobs created are lower to mid-level production workers, with the mid-level being coveted, but the lower level not so much.

So, we have middle managers managing either an untrained or non-existent workforce. This cannot be sustained long term, as the inefficiency will cause a structure change.

So why don't people want to take these lower-level jobs? Where are the people to take these jobs?

Pay, attractiveness of the job, and a smaller workforce are the simple answers. Some more complex answers would involve why there is not an attractiveness to making some money? Why is the workforce smaller? Let's tackle the latter question, which will lead into the former.

The workforce is aging and has been for some years, but it is more pronounced now and in the near future. The Baby Boomer generation is coming to retirement. And Baby Boomers are one of the largest generations. This is creating a large amount of jobs that need to be filled. Some of these openings will be executive and mid-level jobs, but a fair amount will be the lower-level jobs. Why? Well, the next generation, Generation X, would move up to the executive and mid-level jobs, in theory. Gen X is one of the smaller generations as a percentage of the population, so there will still be a good number of executive and mid-level jobs that will go unfilled by Gen X.

And what about Generation Y/Millennials?

While Generation Y/Millennials also make up a large percentage of the population, they typically have a different set of values and criteria for their life choices.

Unlike the previous two generations, who placed a higher value on money, Millennials' main priority is looking for a better balance between work and personal life.

This is not to say the other generations did not value a personal life; they just placed a higher value on providing for themselves and their families.

They wanted their freedom.

For Millennials, they put money on a much lower pedestal. They want to enjoy what they do and to be able to spend time with their friends and family. They want to take vacations. Low-level jobs will not satisfy these values. Some may call this privileged, but it really is not. They do not want to settle.

Consequently, we have a shortage of resources to fill the need to produce more. How does this have anything to do with the future or technology? Well, companies will have to be more creative to make themselves efficient, and what they will do is take some of these low-level job functions and automate them. They will take repetitive jobs and use Automatic Guided Vehicles (AGVs) and robotic tooling. AGVs are also taking over non-value-added jobs, such as transporting material long distances within a facility or between facilities.

This would free up those human resources to do more satisfying, complex functions. This would also decrease the need for as many humans as resources. Robots would take over many of the repetitive functions that would require more manipulation skills, having the same effect as AGVs.

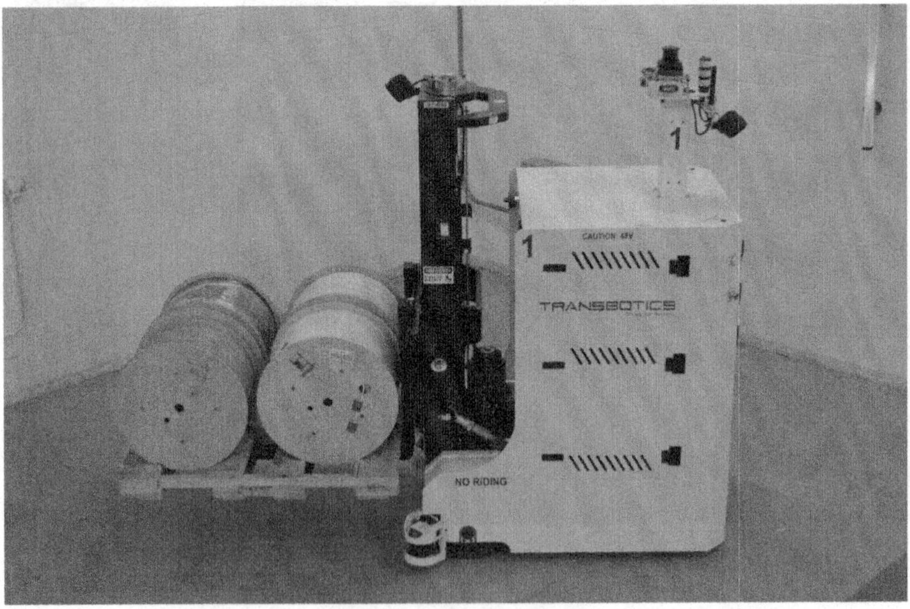

AGVs and robots have been around for a long time, so nothing new there.

Work efficiency has been around for even longer.

So, what's new? What's innovative? Why are you reading this?

Well, what if you need to have manipulation and move long distances because robots need to be secured to a base? Mobile robots. Yes, they have been around for a few years now, but never on a large scale, and not with anything more than a pick-and-place functionality. We will see more offerings with welding, riveting, and the like, as well as larger robotic arms.

In the next year, there will be more people going the Reese's Peanut Butter route. If you don't get this reference, then check YouTube...or is it Vimeo?

As to why you are reading this, it's because you wanted to see where this all ended up. And as for my many questions, I hope you're glad I asked them! ~

Jayesh Mehta is Transbotics' Marketing Manager and co-author of *A Rogue's Guide to Acquisition: Principles from the Final Frontier.*

THE FUTURIST INSTITUTE

FI THE FUTURIST INSTITUTE

The Futurist Institute was founded in 2016 to help analysts, executives, and professionals incorporate new and emerging technology risk into their strategic planning. The Futurist Institute confers the Futurist and Long-Term Analyst™ (FLTA) designation and helps analysts become Certified Futurists™. Our courses have been approved for continuing education hours by the Certified Financial Planner Board of Standards (CFP Board), Global Association of Risk Professionals (GARP), and National Association of Certified Valuators and Analysts (NACVA).

Current Courses

The Future of Work
The Future of Data
The Future of Energy
The Future of Finance
The Future of Healthcare
The Future of Leadership
The Future of Transportation
Futurist Fundamentals
Quantum Computing

Visit The Futurist Institute:

www.futuristinstitute.org

THE FUTURIST INSTITUTE

Visit The Futurist Institute

www.futuristinstitute.org

ABOUT THE EDITOR

Jason Schenker is the President of Prestige Economics and the world's top-ranked financial market futurist. Bloomberg News has ranked Mr. Schenker one of the most accurate forecasters in the world in 42 different categories since 2011, including #1 in the world in 25 categories for his forecasts of the Euro, the Pound, the Swiss Franc, the Chinese RMB, crude oil prices, natural gas prices, gold prices, industrial metals prices, agricultural prices, US nonfarm payrolls, and US new home sales.

Mr. Schenker has written nine books, four of which have been #1 Best Sellers on Amazon: *Commodity Prices 101*, *Recession-Proof*, *Electing Recess*ion, and *Jobs for Robots*. He also edited the #1 Best Seller *The Robot and Automation Almanac - 2018*. Mr. Schenker is also a columnist for *Bloomberg Opinion*. Mr. Schenker has appeared as a guest and guest host on Bloomberg Television as well as a guest on CNBC. He is frequently quoted in the press, including *The Wall Street Journal, The New York Times*, and *The Financial Times.*

Prior to founding Prestige Economics, Mr. Schenker worked for McKinsey & Company as a Risk Specialist, where he directed trading and risk initiatives on six continents. Before joining McKinsey, Mr. Schenker worked for Wachovia as an Economist.

Mr. Schenker holds a Master's in Applied Economics from UNC Greensboro, a Master's in Negotiation from CSU Dominguez Hills, a Master's in German from UNC Chapel Hill, and a Bachelor's with distinction in History and German from The University of Virginia. He also holds a certificate in FinTech from MIT, an executive certificate in Supply Chain Management from MIT, a graduate certificate in Professional Development from UNC, an executive certificate in Negotiation from Harvard Law School, and a certificate in Cybersecurity with NACD and Carnegie Mellon University. Mr. Schenker holds the professional designations CMT® (Chartered Market Technician), CVA® (Certified Valuation Analyst), ERP® (Energy Risk Professional), CFP® (Certified Financial Planner), and FLTA™ (Certified Futurist and Long-Term Analyst). Mr. Schenker is also an instructor for LinkedIn Learning. His courses include Financial Risk Management, Recession-Proof Strategies, Audit & Due Diligence, and a weekly Economic Indicator Series.

Mr. Schenker is a member of the Texas Business Leadership Council, the only CEO-based public policy research organization in Texas, with a limited membership of 100 CEOs and Presidents. He is also a member of the 2018 Director class of the Texas Lyceum, a non-partisan, non-profit that fosters business and policy dialogue on important US and Texas issues.

Mr. Schenker is an active executive in FinTech as a member of the Central Texas Angel Network and as the Executive Director of the Texas Blockchain Association. He is also a member of the National Association of Corporate Directors as well as an NACD Board Governance Fellow.

In October 2016, Mr. Schenker founded The Futurist Institute to help analysts, strategists, executives, and professionals become futurists through a formal training and certification program.

Mr. Schenker was ranked one of the top 100 most influential financial advisors in the world by Investopedia in 2018.

About Jason Schenker:

www.jasonschenker.com

About The Futurist Institute:

www.futuristinstitute.org

ABOUT THE PUBLISHER

Prestige Professional Publishing LLC was founded in 2011 to produce insightful and timely professional reference books. We are registered with the Library of Congress.

Published Titles

Be The Shredder, Not The Shred

Commodity Prices 101

Electing Recession

Jobs for Robots

Midterm Economics

The Promise of Blockchain

Quantum: Computing Nouveau

Robot-Proof Yourself

The Robot and Automation Almanac — 2018

The Robot and Automation Almanac — 2019

Future Titles

A Gentle Introduction to Audit and Due Diligence

Spikes: Growth Hacking Leadership

JOBS FOR ROBOTS

Jobs for Robots provides an in-depth look at the future of automation and robots, with a focus on the opportunities as well as the risks ahead. Job creation in coming years will be extremely strong for the kind of workers that do not require payroll taxes, health care, or vacation: robots. *Jobs for Robots* was published in February 2017. This book has been a #1 Best Seller on Amazon.

ROBOT-PROOF YOURSELF

Robot-Proof Yourself offers a number of practical professional recommendations for how to be robot-proof in the coming era of professional, economic, and financial disruptions. Robots and automation are set to advance, but individuals have a chance to benefit from the coming changes. *Robot-Proof Yourself* was released in December 2017.

DISCLAIMER

FROM THE PUBLISHER

The following disclaimer applies to any content in this book:

This book is commentary intended for general information use only and is not investment advice. Prestige Professional Publishing LLC does not make recommendations on any specific or general investments, investment types, asset classes, unregulated markets, specific equities, bonds, or other investment vehicles. Prestige Professional Publishing LLC does not guarantee the completeness or accuracy of analyses and statements in this book, nor does Prestige Professional Publishing LLC assume any liability for any losses that may result from the reliance by any person or entity on this information. Opinions, forecasts, and information are subject to change without notice. This book does not represent a solicitation or offer of financial or advisory services or products; this book is only market commentary intended and written for general information use only. This book does not constitute investment advice.

DISCLAIMER

FROM THE PUBLISHER

DISCLAIMER

FROM THE FUTURIST INSTITUTE

The following disclaimer applies to any content in this book:

This book is commentary intended for general information use only and is not investment advice. The Futurist Institute does not make recommendations on any specific or general investments, investment types, asset classes, non-regulated markets, specific equities, bonds, or other investment vehicles. The Futurist Institute does not guarantee the completeness or accuracy of analyses and statements in this book, nor does The Futurist Institute assume any liability for any losses that may result from the reliance by any person or entity on this information. Opinions, forecasts, and information are subject to change without notice. This book does not represent a solicitation or offer of financial or advisory services or products, and are market commentary intended and written for general information use only. This book does not constitute investment advice.

FI THE FUTURIST INSTITUTE

THE ROBOT AND AUTOMATION ALMANAC - 2019

COMPILED BY THE FUTURIST INSTITUTE

EDITED BY JASON SCHENKER

ISBN: 978-1-946197-14-6 *Paperback*
978-1-946197-15-3 *Ebook*

PRESTIGE
PROFESSIONAL PUBLISHING

Made in the USA
Monee, IL
13 September 2026

62976627R00105